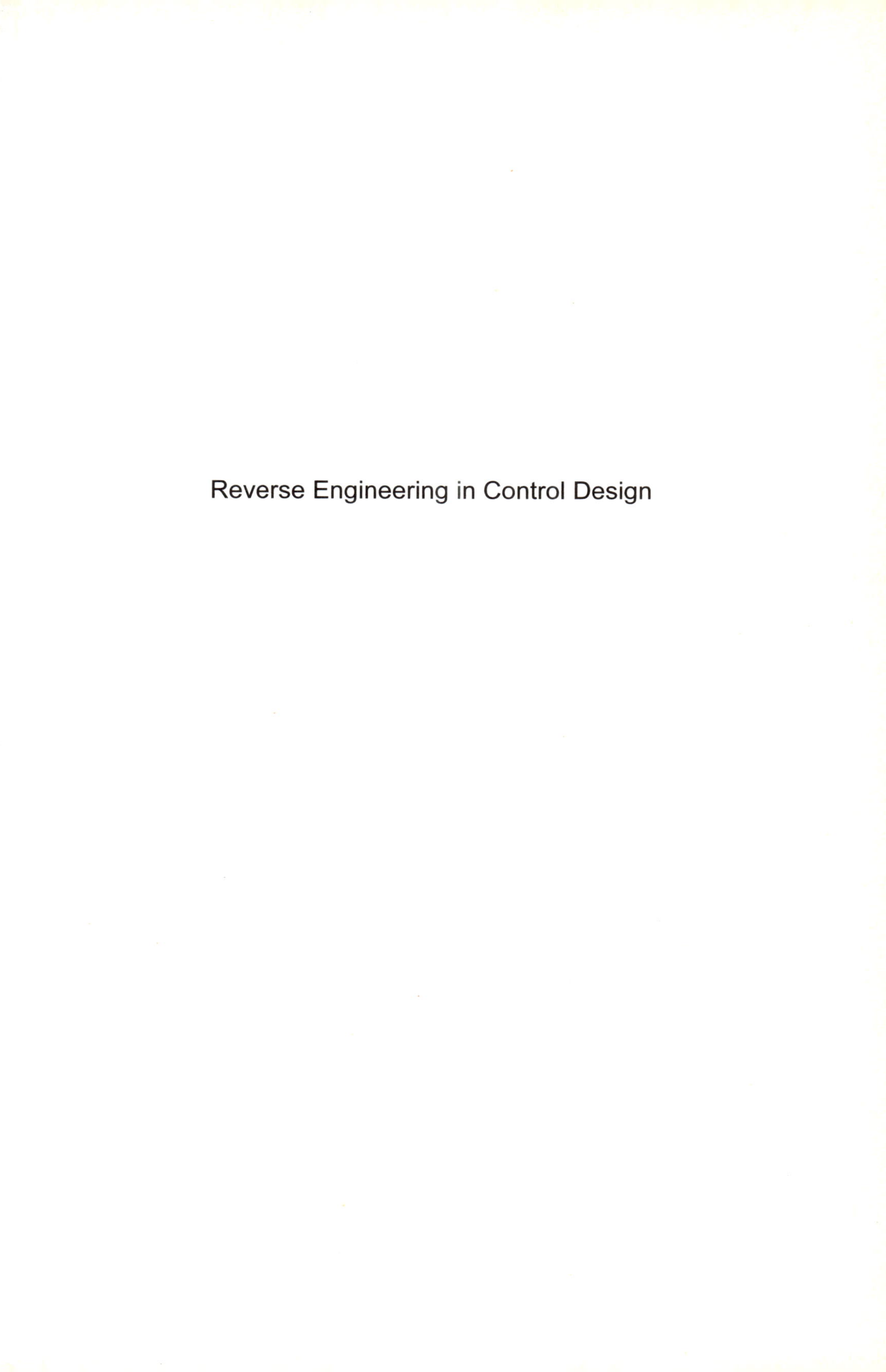

Reverse Engineering in Control Design

I would like to thank all my colleagues from the Institut Supérieur de l'Aéronautique et de l'Espace (ISAE) and from the Department of System Control and Flight Dynamics, ONERA (the French Aerospace Lab) for their help and support in this research. I would also like to thank all my PhD students, present and past, and more particularly to Olivier Voinot, Fabien Delmond, Nicolas Fezans and Nicolas Guy whose research works have fed this book.

Daniel Alazard

FOCUS SERIES IN AUTOMATION AND CONTROL – INDUSTRIAL ENGINEERING

Series Editor Bernard Dubuisson

Reverse Engineering in Control Design

Daniel Alazard

WILEY

First published 2013 in Great Britain and the United States by ISTE Ltd and John Wiley & Sons, Inc.

ISTE Ltd
27-37 St George's Road
London SW19 4EU
UK

www.iste.co.uk

John Wiley & Sons, Inc.
111 River Street
Hoboken, NJ 07030
USA

www.wiley.com

Library of Congress Control Number: 2012955967

British Library Cataloguing-in-Publication Data
A CIP record for this book is available from the British Library
ISSN: 2051-2481 (Print)
ISSN: 2051-249X (Online)
ISBN: 978-1-84821-523-8

Printed and bound in Great Britain by CPI Group (UK) Ltd., Croydon, Surrey CR0 4YY

Table of Contents

Nomenclature

Following notations are used throughout this book.

A^*	A conjugated and transposed
A^{T}	A transposed
A^+	Moore–Penrose pseudo-inverse of matrix A
$A^\perp$	Orthonormal basis for the null space of A
spec(A)	set of eigenvalues for a square matrix A
Ker(A)	null space of matrix A
diag(λ_i)	diagonal $n \times n$ matrix built from $\lambda_i,\ i = 1 \cdots n$
I_n	$n \times n$ identity matrix
$\mathbb{R}$	set of real numbers
$\mathbb{C}$	set of complex numbers
i	$\sqrt{-1}$
$\|c\|$	magnitude of the complex number c
$\dot{x}$	time derivation ($\dot{x} = dx/dt$)
s	Laplace variable
LQG	linear-quadratic-Gaussian
LFT	linear fractional transformation
$F_l(P, K)$	lower linear fractional transformation of P and K
$P_{w_i \to z_i}(s)$	the transfer from w_i to z_i in the plant $P(s)$
$\\|G(s)\\|_2$	H_2 norm of the stable system $G(s)$
$\\|G(s)\\|_\infty$	H_∞ norm of the stable system $G(s)$
$\sigma_{min}(C)$	minimal singular value of the complex matrix C
$\sigma_{max}(C)$	maximal singular value of the complex matrix C
$G(s) := \left[\begin{array}{c\|c} A & B \\ \hline C & D \end{array}\right]$	shorthand for $G(s) = C(sI - A)^{-1}B + D$

Introduction

The first objective of this book is to provide a general solution to the *inverse* H_∞ and H_2 *optimal control problems* and to show how such a solution can be used to design controllers in a *reverse engineering* approach. Given an initial controller and a given plant, the solution to the *inverse* H_∞/H_2 *optimal control problem* is a standard H_∞ control problem, that is the two input ports–two output ports standard problem commonly used in the H_∞ design framework, whose unique H_∞ or H_2 optimal controller is the given controller. This solution will be called the *cross standard form* (CSF). The *reverse engineering* approach consists of applying the CSF to a given controller in order to set up a standard control problem that can be completed to handle H_2 or H_∞ frequency-domain specifications. It will be shown that such an approach is quite attractive to mix various control design methods and to cope with various kinds of control specifications. Thus, *reverse engineering* is a suitable alternative to multi-objective control design that is still an open problem.

The second objective, which is strongly linked to the first objective, concerns the *observer-based realization* of controllers for implementation purposes. Full-order H_∞ unstructured controllers are well-known to raise implementation problems such as:

– How the controller states can be initialized.

– How various controllers designed at various operating conditions can be switched or gain-scheduled; and so on.

When the model of a plant is described by a state space representation where the states have a physical meaning (and physical units: this is in particular the case in the field of mechanical engineering where the plant state vector is composed of the displacements along the various degrees of freedom and their time derivatives), an observer-based controller has a physical structure. Its state is an estimate of the plant state and has the same physical units. All the gains (state-feedback gains and state-estimator gains) have also a physical unit. This can reduce implementation problems in a significant way. It will be shown how an observer-based realization of a given controller for a given plant can be computed and implemented. Then, the link between *observer-based realization* and *reverse engineering* is straightforward: the observer-based realization of a controller allows a simple solution to the inverse optimal control problem to be proposed.

The third objective of this book concerns *reverse engineering* for mechanical systems and initial controllers designed to meet basic performance specifications, that is a prescribed second-order behavior for each degree of freedom and dynamic decoupling of degrees of freedom. The CSF, although a general solution to the inverse H_∞ optimal control problem, leads to a standard problem where the closed-loop performance index cannot be always directly compared with other performance indexes. To solve this problem, a new standard H_∞ problem weighting the *acceleration sensitivity function* is proposed as a starting point for the *reverse engineering* approach.

These three objectives are described in detail in Chapters 1–3. In Chapter 1, we present the procedure to compute the observer-based realization of a given controller and a given model. The application of this procedure to a very simple model of a launcher is proposed to illustrate the importance of observer-based controllers for gain-scheduling, controller switching, state and disturbance monitoring, and reference input tracking. In Chapter 2, the CSF is presented and also applied to the same academic example: a low-order controller is improved to fulfill a template on its frequency-domain response and to recover stability margins once the actuator dynamics is taken into account. Chapter 3 discusses reverse engineering for the particular class of mechanical systems. The extension of these results to the discrete-time case is given in the appendix. Concluding remarks and future works are proposed in the last chapter.

In the three chapters, academic applications are proposed to illustrate the various basic principles. These applications have been chosen for their pedagogic contents: demo files and embedded Matlab® functions can be downloaded from http://personnel.isae.fr/daniel-alazard/matlab-packages. Readers can run these illustrations on their personnel computer. More complex and more realistic applications related to this book are referenced for the reader who wishes to go further. The reader is also advised to read first the preliminary methodological example proposed in Appendix 1 to better understand the motivations of this book.

This book is aimed mainly for postgraduate students and control designers with a solid background in automatic control, mainly in:

– state-space representation of multivariable systems;

– linear–quadratic–gaussian (LQG), H_2 and H_∞ syntheses.

1

Observer-based Realization of a Given Controller

1.1. Introduction

Observer-based controllers (for instance, Linear Quadratic Gaussian (LQG) controllers) are quite interesting for different practical reasons and from the implementation point of view. Probably the key advantage of these controller structures lies in the fact that the controller states are meaningful variables as estimates of the physical plant states. It follows that the controller states can be used to monitor (online or offline) the performance of the system. Such a meaningful state also allows us to initialize the state of the controller or to update the controller state during control mode switching. Note that this simple property does not hold for general controllers with state-space description:

$$\begin{cases} \dot{x}_K = A_K x_K + B_K y \\ u = C_K x_K + D_K y. \end{cases} \qquad [1.1]$$

Another well-appreciated advantage comes from the ease of implementation of observer-based controllers. In addition to the plant data, only two static gains (the state-feedback and the state-estimator) define the entire controller dynamics. In return, this facilitates the construction of gain-scheduled or interpolated controllers. Indeed, assuming the plant model is available in real time, observer-based controllers will only require the storage of these two static gains of lower dimensions instead of the huge set of numerical data in [1.1] to update the controller dynamics at each sample of time. Note that if we are using an interpolating procedure to update the controller dynamics, the general representation in [1.1] is highly questionable from an implementation viewpoint and in many cases will lead to an insuperable computational effort. This was in our opinion a major impediment for a widespread use of modern control techniques, such as H_∞ and μ syntheses in realistic applications and particularly for problems necessitating the real-time adjustment of the controller gains. These approaches produce high-order controllers expressed under a meaningless state-space realization. Note also that this last point is relevant if a controller reduction has been performed after the design.

To overcome this problem a general procedure is proposed in this chapter to compute an observer-based realizations for an arbitrary given controller and a given plant (for both continuous and discrete time cases). Independently of the solver used for the control design, such a procedure allows us to provide a realization with a meaningful state vector. In [ALA 01] and [CUM 04], it is shown that observer-based realization are also convenient to isolate high level-tuning parameters (potentiometers) in a complex control law. As the observer-based realization exploits the model of the plant, we can also guess that such a realization is very convenient to update the controller to a change in the model or to build a parameter-dependent controller $K(s, \theta)$ from a parameter-dependent model $G(s, \theta)$.

Among other potential advantages of observer-based realization, we would like to point out the possibility to handle actuator saturation constraints by exploiting this information into the prediction equation. Since this matter is not covered in this book, the reader is referred to [TAR 97] and references therein for more details. More theoretical discussions on the implementation of gain-scheduled controllers which use the plant nonlinearity model are given in [LAW 95] and [KAM 95].

The practical solution to handle non-stationary problems (such as the launch vehicle control design during atmospheric flight) or nonlinear problems consists of designing a family of controllers at various flight instants or various flight conditions and then in interpolating (gain-scheduling) these various controllers. It is well-known that the non-stationary behavior of interpolated control laws depends strongly on the controller realizations that are interpolated. Observer-based realizations are very attractive from the gain scheduling point of view [STI 99, PEL 00]. The main reason is that the controller states are consistent and have physical units if the model upon which the observer-based realization is built has physical states. Then, observer-based realizations of given controllers is a good alternative to provide gain-scheduled controllers.

In this chapter, we briefly recall central ideas behind the Youla parametrization and show how it can be used to find the state-estimator-state-feedback structure of an arbitrary controller associated with a given plant. Then, some illustrative examples, with the associated Matlab® sequence, are proposed to highlight the interest of observer-based structures to solve practical problems: gain-scheduling, controller switching, state and disturbance monitoring and reference input tracking.

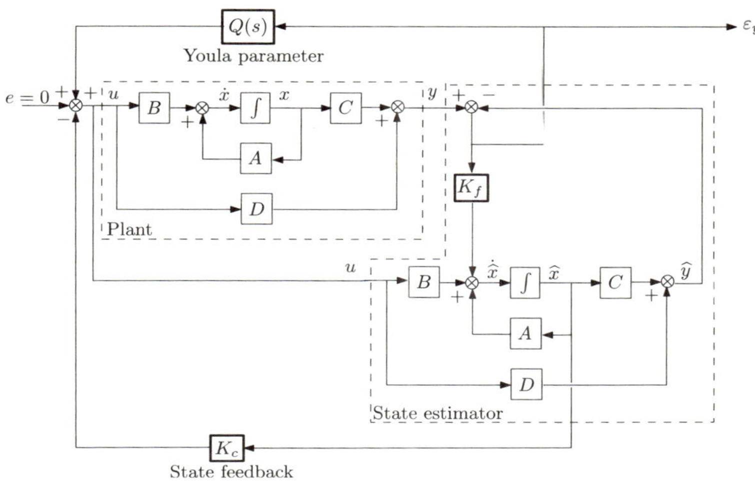

Figure 1.1. *Observer-based structure and Youla parametrization*

1.2. Principle

Consider the stabilizable and detectable nth-order plant $G_0(s)$ (m inputs and p outputs) with a minimal state-space realization:

$$\begin{cases} \dot{x} = Ax + Bu, \\ y = Cx + Du \end{cases} \quad \text{also noted:} \quad \begin{bmatrix} \dot{x} \\ \hline y \end{bmatrix} = \left[\begin{array}{c|c} A & B \\ \hline C & D \end{array}\right] \begin{bmatrix} x \\ \hline u \end{bmatrix}. \qquad [1.2]$$

The so-called Youla parametrization of all stabilizing controllers built on the general observer-based structure is depicted in Figure 1.1, where K_c, K_f and $Q(s)$ are the state-feedback gain, the state-estimator gain and the Youla parameter, respectively. The controller $K(s)$ associated with this structure is easily shown to have the following state-space description:

$$\begin{cases} \dot{\widehat{x}} = A\widehat{x} + Bu + K_f(y - C\widehat{x} - Du) \\ \dot{x_Q} = A_Q x_Q + B_Q(y - C\widehat{x} - Du) \\ u = -K_c\widehat{x} + C_Q x_Q + D_Q(y - C\widehat{x} - Du) \end{cases} \qquad [1.3]$$

where A_Q, B_Q, C_Q and D_Q are the four matrices of the state-space representation of $Q(s)$ associated with the state variable x_Q. Hereafter, $\widehat{x}$ denotes an estimate of the plant state x. Such a state-space realization can also be seen as the lower Linear Fractional Transformation (LFT) of $Y(s)$ and $Q(s)$ where $Y(s)$ is defined by:

$$\begin{bmatrix} \dot{\widehat{x}} \\ \hline u \\ \varepsilon_y \end{bmatrix} = \left[\begin{array}{c|cc} A - BK_c - K_fC + K_fDK_c & K_f & B - K_fD \\ \hline -K_c & 0 & I_m \\ -C + DK_c & I_p & -D \end{array}\right] \begin{bmatrix} \widehat{x} \\ \hline y \\ e \end{bmatrix}. \qquad [1.4]$$

That is:

$$K(s) = F_l(Y(s), Q(s)). \qquad [1.5]$$

The principle of the Youla parametrization of all the controllers stabilizing the plant G_0 is based on the fact that the closed-loop transfer function between the input e and the innovation $\varepsilon_y = y - C\widehat{x} - Du$ is null (see [LUE 71], for instance). As a consequence, changing $Q(s)$ leads to various controllers but the closed-loop transfer function remains unaffected. It is readily shown that this closed-loop transfer function can be represented by the state-space form [1.6] involving the estimation error $\varepsilon_x = x - \widehat{x}$:

$$\begin{aligned} \begin{bmatrix} \dot{x} \\ \dot{x_Q} \\ \dot{\varepsilon_x} \end{bmatrix} &= \begin{bmatrix} A - BK_c & BC_Q & BK_c + BD_QC \\ 0 & A_Q & B_QC \\ 0 & 0 & A - K_fC \end{bmatrix} \begin{bmatrix} x \\ x_Q \\ \varepsilon_x \end{bmatrix} + \begin{bmatrix} B \\ 0 \\ 0 \end{bmatrix} e \\ \varepsilon_y &= \begin{bmatrix} 0 & 0 & C \end{bmatrix} \begin{bmatrix} x \\ x_Q \\ \varepsilon_x \end{bmatrix} \end{aligned} \qquad [1.6]$$

From this representation, the separation principle appears clearly and can be stated in the following terms:

– the closed-loop eigenvalues can be separated into n state-feedback eigenvalues (spec$(A - BK_c)$), n state-estimator eigenvalues (spec$(A - K_fC)$) and the Youla parameter eigenvalues (spec(A_Q)),

– the state-estimator eigenvalues and the Youla parameter eigenvalues are uncontrollable by e;

– the state-feedback eigenvalues and the Youla parameter eigenvalues are unobservable from ε_y. The transfer function from e to ε_y always vanishes.

Thus, the closed-loop system is stable if and only if matrices $(A - BK_c)$, $(A - K_fC)$ and A_Q are Hurwitz (all the eigenvalues lie in the open left half plane). Then, changing $Q(s)$ by any stable transfer allows us to parametrize the set of all controllers stabilizing the plant $G_0(s)$.

Now let us consider a stabilizing n_Kth-order controller $K_0(s)$ with minimal state-space realization:

$$\begin{cases} \dot{x_K} = A_K x_K + B_K y & [1.7a] \\ u = C_K x_K + D_K y, & [1.7b] \end{cases}$$

the objective is to find its Youla parametrization associated with the plant realization (equation [1.2]), that is the set of parameters $\{K_c,\ K_f,\ Q(s)\}$ such that the state-space realization (1.3) is a *minimal* observer-based realization of $K_0(s)$. The minimality of the observer-based realization developed in this chapter must be emphasized by comparison with the more general parametrization proposed in [BOY 91] and [ZHO 96], for instance. The Youla parametrization of the controller proposed in these references is based on a co-prime factorization of the plant and leads to a realization whose order could be $2n + n_K$ and where the minimality is not guaranteed.

In the sequel, the following notations will be used:

$$\boxed{J_m = (I_m - D_K D)^{-1} \quad \text{and} \quad J_p = (I_p - D D_K)^{-1},} \qquad [1.8]$$

with the following properties:

– $J_m D_K = D_K J_p$, $J_p D = D J_m$,

– $I_m + D_K D J_m = J_m$, $I_p + J_p D D_K = J_p$.

We first express the controller state equation [1.7a] as an Luenberger observer of the variable $z = Tx$. So, we will denote:

$$x_K = \widehat{z} \qquad [1.9]$$

According to Luenberger's formulation [LUE 71], this problem can be stated as the search of:

$$T \in \mathbb{R}^{n_K \times n},\ F \in \mathbb{R}^{n_K \times n_K},\ G \in \mathbb{R}^{n_K \times p},$$

such that

$$\dot{\widehat{z}} = F\widehat{z} + G(y - Du) + TBu \qquad [1.10]$$

is an (asymptotic) observer of the variable z, that is $z - \widehat{z}$ vanishes as t goes to infinity with an estimation dynamics given by F. Luenberger has shown that the constraints:

$$TA - FT = GC, \text{ and } F \text{ stable,} \qquad [1.11]$$

ensure that this holds true. Then, with the output equation [1.7b], the state-space representation of the controller reads:

$$\begin{cases} \dot{\widehat{z}} = \Big(F + (TB - GD)C_K\Big)\widehat{z} + \Big(G(I_p - DD_K) + TBD_K\Big)y & \text{(a)} \\ u = C_K\widehat{z} + D_K y & \text{(b)} \end{cases}. \qquad [1.12]$$

The structure of the controller is then depicted in Figure 1.2.

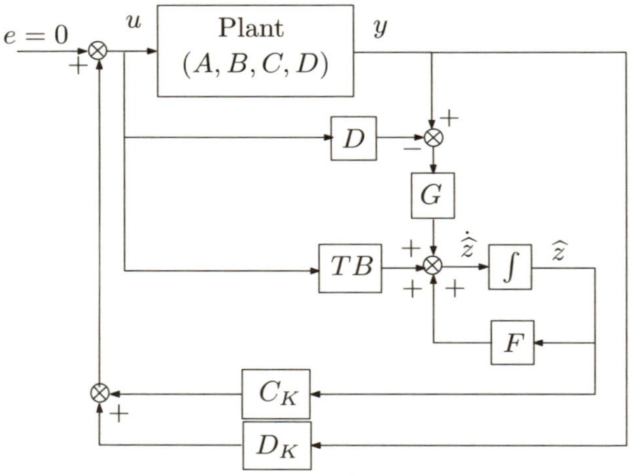

Figure 1.2. *Controller structure as an observer of* $z = Tx$

With [1.9], the identification of [1.12] and [1.7a] leads to the algebraic relations:

$$G = (B_K - TBD_K)J_p. \qquad [1.13]$$

$$F = A_K + (B_K D - TB)J_m C_K. \qquad [1.14]$$

These equations with [1.11] guarantee that we are dealing with an observer-based controller. Note that the stability of F (equation [1.11]) is secured whenever the original controller [1.7a] is stabilizing. Indeed from [1.2] and [1.12], a closed-loop state-space realization reads:

$$\begin{bmatrix} \dot{x} \\ \dot{\widehat{z}} \end{bmatrix} = \begin{bmatrix} A + BJ_m D_K C & BJ_m C_K \\ GC + TBJ_m D_K C & F + TBJ_m C_K \end{bmatrix} \begin{bmatrix} x \\ \widehat{z} \end{bmatrix}. \qquad [1.15]$$

Let us consider the change of state coordinates involving the estimation error $\varepsilon_z = z - \widehat{z}$:

$$\begin{bmatrix} x \\ \widehat{z} \end{bmatrix} = \mathcal{M} \begin{bmatrix} x \\ \varepsilon_z \end{bmatrix} \quad \text{with } \mathcal{M} = \begin{bmatrix} I_n & 0 \\ T & -I_{n_K} \end{bmatrix} \quad \text{and } \mathcal{M}^{-1} = \mathcal{M}. \qquad [1.16]$$

The new state-space realization highlights the separation principle:

$$\begin{bmatrix} \dot{x} \\ \dot{\varepsilon}_z \end{bmatrix} = \begin{bmatrix} A + BJ_m(D_K C + C_K T) & -BJ_m C_K \\ 0 & F \end{bmatrix} \begin{bmatrix} x \\ \varepsilon_z \end{bmatrix}. \qquad [1.17]$$

So the set of $n + n_K$ closed-loop eigenvalues include the n_K eigenvalues of F. Therefore, F, that is the estimation dynamics, is stable if the initial controller is stabilizing.

Substituting F and G from [1.14] and [1.13] in the first relation [1.11], we get:

$$\boxed{(A_K + B_K D J_m C_K)T - T(A + BJ_m D_K C) - TBJ_m C_K T + B_K J_p C = 0.} \qquad [1.18]$$

So, the problem is reduced to solve in T the generalized non-symmetric and rectangular Riccati equation [1.18] and, in the following, to compute F and G using [1.14] and [1.13], respectively.

Equation [1.18] can also be reformulated as:

$$[-T \quad I] A_{cl} \begin{bmatrix} I \\ T \end{bmatrix} = 0, \qquad [1.19]$$

where the characteristic matrix A_{cl} associated with the Riccati equation [1.18] is nothing else than the closed-loop system matrix:

$$A_{cl} := \begin{bmatrix} A + BJ_m D_K C & BJ_m C_K \\ B_K J_p C & A_K + B_K D J_m C_K \end{bmatrix}. \qquad [1.20]$$

The Riccati equation [1.18] can then be solved by standard invariant subspace techniques that consist of:

– finding an n-dimensional invariant subspace $\mathcal{S}: = Range(U)$ of the closed-loop system matrix A_{cl}, that is

$$A_{cl}U = U\Lambda. \qquad [1.21]$$

This subspace is associated with a set of n eigenvalues, spec(Λ), among the $n+n_K$ eigenvalues of A_{cl}. Such subspaces are easily computed using Schur factorizations or eigenvalue decompositions of the matrix A_{cl}. See [GOL 96] for more details:

– Partitioning the vectors U which span this subspace conformably to the partitioning in [1.20]:

$$U = \begin{bmatrix} U_1 \\ U_2 \end{bmatrix}, \quad U_1 \in \mathbb{R}^{n\times n}. \qquad [1.22]$$

– Computing the solution:

$$\boxed{T = U_2 U_1^{-1}.} \qquad [1.23]$$

Narasimhamurthi and Wu [NAR 77] have shown that the existence of a solution T satisfying [1.18] is guaranteed whenever the eigenvalues of the Hamiltonian matrix A_{cl} are distinct. In proposition 1.2, a necessary condition is given for the existence of a solution T. In the general case, however, there are finitely many admissible subspaces $\mathcal{S}$ and thus many solutions. Each solution corresponds to a particular choice of n eigenvalues among the set of closed-loop eigenvalues of A_{cl}.

Then, given an nth-order plant and an n_Kth-order controller, we can compute the linear combination $T_{n_K\times n}x$ of the plant states which is estimated by the controller state. An analogous result is also discussed by Bender and Fowell [BEN 86].

The reader can download from: http://personnel.isae.fr/daniel-alazard /matlab-packages a toolbox (`bib_obr`) with an interactive Matlab® function `cor2tfg` to compute the matrices T, F and G from a given controller $K_0(s)$ and a given plant $G_0(s)$ (the `help` of this function can be found in section A4.1, Appendix 4).

1.3. A first illustration

The model of a launcher between the angle of attack α and the thruster deflection δ can be roughly approximated by the second-order transfer function (i.e. $n = 2$):

$$G_0(s) = \frac{1}{s^2 - 1}$$

or the state-space realization ($x = [\alpha \quad \dot{\alpha}]^{\mathrm{T}}$):

$$\left[\begin{array}{c} \dot{\alpha} \\ \ddot{\alpha} \\ \hline \alpha \end{array}\right] = \left[\begin{array}{cc|c} 0 & 1 & 0 \\ 1 & 0 & 1 \\ \hline 1 & 0 & 0 \end{array}\right] \left[\begin{array}{c} \alpha \\ \dot{\alpha} \\ \hline \delta \end{array}\right]. \qquad [1.24]$$

Let us consider the following first-order stabilizing controller ($n_K = 1$) connected to the plant with positive feedback (see Figure 1.3):

$$K_0(s) = \frac{-23s - 32}{s + 12}.$$

A state-space realization of this controller reads:

$$\left[\begin{array}{c} \dot{x_K} \\ \hline \delta \end{array}\right] = \left[\begin{array}{c|c} -12 & 4 \\ \hline 61 & -23 \end{array}\right] \left[\begin{array}{c} x_K \\ \hline \alpha \end{array}\right]. \qquad [1.25]$$

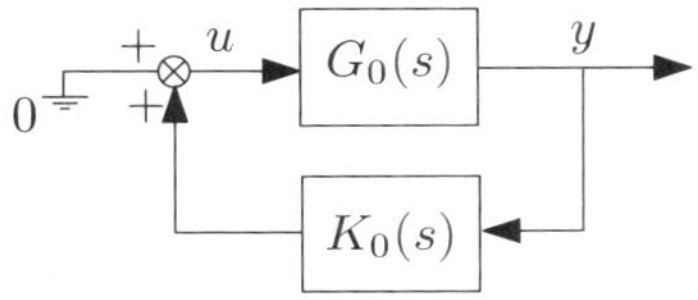

Figure 1.3. *Positive feedback connection of plant $G_0(s)$ and controller $K_0(s)$*

The closed-loop dynamic matrix (equation [1.20]) reads in this case:

$$A_{cl} = \begin{bmatrix} 0 & 1 & 0 \\ -22 & 0 & 61 \\ 4 & 0 & -12 \end{bmatrix}$$

and exhibits one real eigenvalue (-10) and two complex auto-conjugate eigenvalues ($-1 \pm i$). To solve in T the Riccati equation [1.19], we have to choose a two-dimensional subspace of A_{cl} associated with two among three closed-loop eigenvalues. It is recommended to select auto-conjugate pairs of eigenvalues to find a real solution T (see also remark 1.4), that is $-1 \pm i$. Then the following Matlab® sequence allows us to determine the solution $T = [0.3279 - 0.0328]$ and matrices G and F using equations [1.13] and [1.14] (note that in this example $D = 0$ then $J_m = I_m$ and $J_p = I_p$):

```
>> G0=ss([0 1;1 0],[0;1],[1 0],0);
>> K0=ss(-12,4,61,-23);
>> CL=feedback(G0,-K0);
>> ACL=CL.a

ACL =
     0     1     0
   -22     0    61
     4     0   -12

>> [U,D]=eig(ACL)

U =
    0.0976              -0.3996 - 0.3996i   -0.3996 + 0.3996i
   -0.9759               0.7992              0.7992
    0.1952              -0.1572 - 0.1310i   -0.1572 + 0.1310i


D =
  -10.0000                    0                    0
         0              -1.0000 + 1.0000i          0
         0                    0              -1.0000 - 1.0000i

>> T=U(3,2:3)*inv(U(1:2,2:3))

T =
    0.3279    -0.0328

>> G=K0.b-T*G0.b*K0.d

G =
    3.2459

>> F=K0.a+(K0.b*G0.d-T*G0.b)*K0.c

F =
  -10.0000
```

To illustrate relation [1.9] that is: x_k is an estimate of the plant variable $z = Tx$ associated with the estimation dynamics F, the response to initial conditions on plant states $x(t = 0) = [1 \quad -1]^{\mathrm{T}}$ are depicted in Figures 1.4 and 1.5:

– Figure 1.4 plots the launcher output α and allows us to appreciate the response time of the closed-loop system: it takes approximately 4 s to bring back the output to 0.

– Figure 1.5 plots the controller state $x_K = \widehat{z}$, the combination of launcher state $z = Tx$ and the estimation error $\varepsilon_z = z - \widehat{z} = Tx - x_K$. It is quite obvious that estimation dynamics (here $F = -10$ rad/s) is quite fast with respect to the closed-loop system response time: the estimation error goes back to zero in less than half a second.

Such figures (Figures 1.4 and 1.5) can be easily obtained with the Matlab® sequence:

```
[y,t,x]=initial(CL,[1 -1 0],6);
figure
plot(t,y,'k-','LineWidth',2);
xlabel('Time (s)');
ylabel('Output y (rd)');
figure
plot(t,x(:,1:2)*T','k-','LineWidth',2);
hold on
plot(t,x(:,3),'k--','LineWidth',2);
plot(t,x(:,1:2)*T'-x(:,3),'g-','LineWidth',2)
xlabel('Time (s)');
ylabel('Controller state x_K and Tx');
legend('Tx','x_K','Tx-x_K');
```

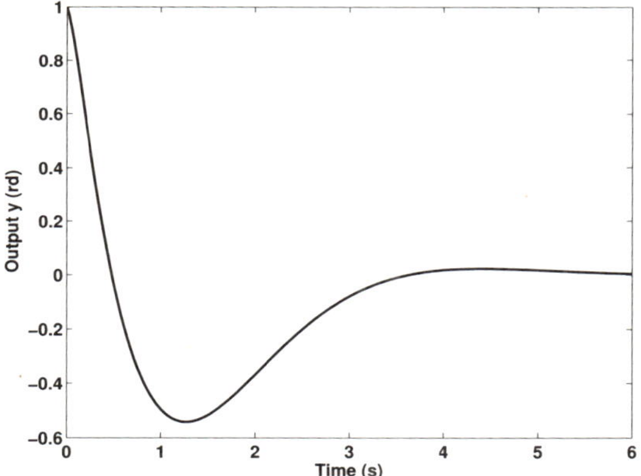

Figure 1.4. *Response of output α to initial conditions on launcher states*

We can also use the interactive function `cor2tfg` to find T, F and G:

```
>> [T,F,G]=cor2tfg(G0,K0)
```

This command line opens two graphics windows (see Figures 1.6):

– the map of closed-loop eigenvalues (marked with $\times$) where the plant open-loop dynamics (marked with $+$) are also marked,

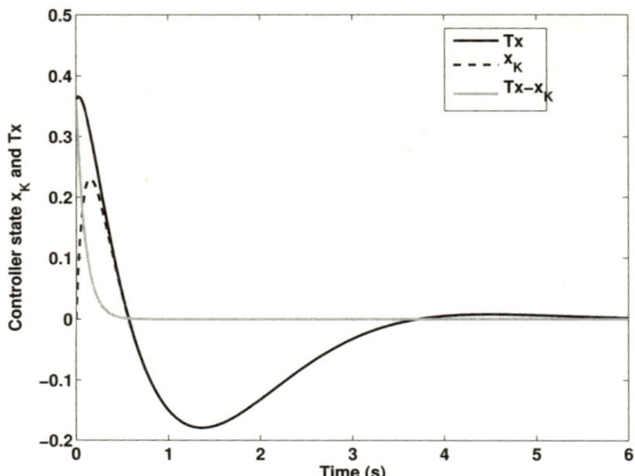

Figure 1.5. *Responses of the controller state x_K, $z = Tx$ and the estimation error $\varepsilon_z = Tx - X_K$ to initial conditions on launcher states*

– a dialog box for the user to assign the current closed-loop eigenvalue.

Answering yes in the dialog box leads to the following result:

```
Selection is completed

T =
     0.3279    -0.0328

F =
   -10.0000

G =
     3.2459
```

1.4. Augmented-order controllers

In this section, we consider the problem where $n_K \geq n$ and find a state-feedback gain K_c, a state-estimator gain K_f and a dynamic Youla parameter $Q(s)$ with order $n_K - n$, such that the observer-based controller structure in Figure 1.1 is equivalent to the original controller [1.7a]. We will assume that T has been computed by the previous technique according to an admissible choice of n eigenvalues among the $n + n_K$ closed-loop eigenvalues. In the following, F and G can be computed from [1.14] and [1.13].

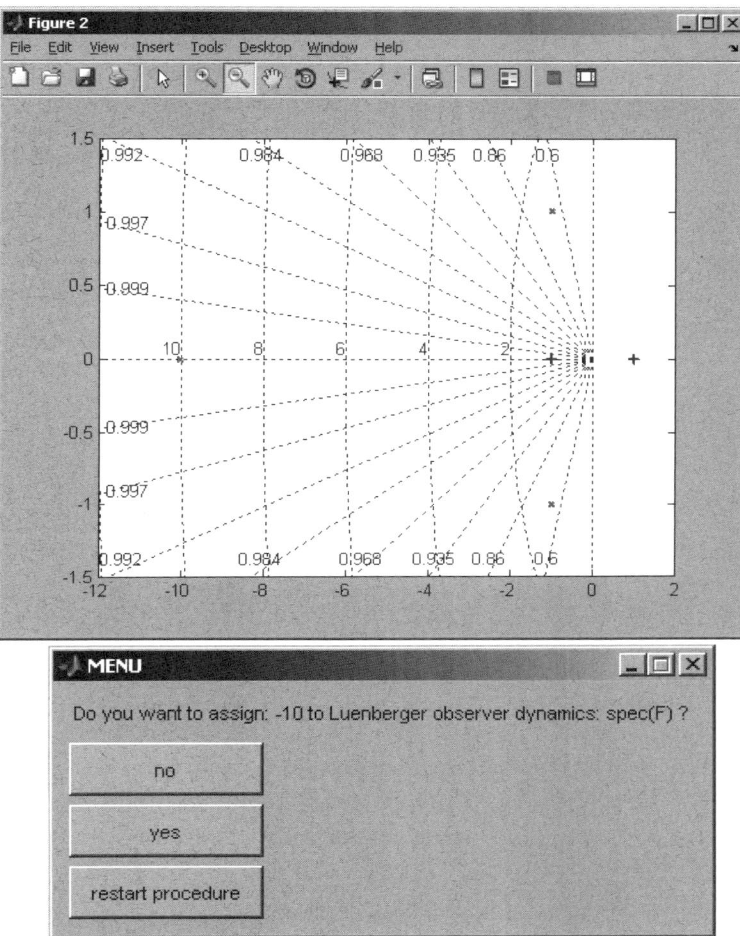

Figure 1.6. *Closed-loop eigenvalues map and dialog box opened by function* `cor2tfg`

Let us consider the Schur decomposition of A_{cl} used to solve in T the Riccati equation [1.18]:

$$A_{cl} = \begin{bmatrix} U_1 & U_3 \\ U_2 & U_4 \end{bmatrix} \begin{bmatrix} \Lambda & * \\ 0 & \Lambda^F \end{bmatrix} \begin{bmatrix} U_1^* & U_2^* \\ U_3^* & U_4^* \end{bmatrix}, \qquad [1.26]$$

where $\begin{bmatrix} U_1 \; U_3 \\ U_2 \; U_4 \end{bmatrix}$ is a unitary $(n + n_K) \times (n + n_K))$ matrix with $U_1 \in \mathbb{C}^{n \times n}$, $U_2 \in \mathbb{C}^{n_K \times n}$, $U_3 \in \mathbb{C}^{n \times n_K}$ and $U_4 \in \mathbb{C}^{n_K \times n_K}$.

From equations [1.15] and [1.17], we can write:

$$\begin{bmatrix} A + BJ_m(D_K C + C_K T) & -BJ_m C_K \\ 0 & F \end{bmatrix} = \begin{bmatrix} I_n & 0 \\ T & -I_{n_K} \end{bmatrix} A_{cl} \begin{bmatrix} I_n & 0 \\ T & -I_{n_K} \end{bmatrix}. \quad [1.27]$$

As $\boxed{T = U_2 U_1^{-1}}$, substituting [1.26] in [1.27] we can derive[1]:

$$F = V \Lambda^F V^{-1} \quad \text{with} \quad \boxed{V = U_2 U_1^{-1} U_3 - U_4}. \quad [1.28]$$

Λ^F is an $n_K \times n_K$ upper triangular matrix that can be decomposed by blocks with block sizes $n_K - n$ and n. The adequate decomposition of V and V^{-1} allows us to write:

$$F = [V_1 \quad V_2] \begin{bmatrix} \Lambda_{11}^F & \Lambda_{12}^F \\ 0 & \Lambda_{22}^F \end{bmatrix} \begin{bmatrix} W_1 \\ W_2 \end{bmatrix} \quad [1.29]$$

with

$$\boxed{V = [\underbrace{V_1}_{n_K - n} \quad \underbrace{V_2}_{n}] \quad \text{and} \quad V^{-1} = \begin{bmatrix} W_1 \\ W_2 \end{bmatrix} \begin{matrix} \}n_K - n \\ \}n \end{matrix}.} \quad [1.30]$$

Let us perform the change of variable:

$$\widehat{z} = \begin{bmatrix} V_1 & V_2 \end{bmatrix} \begin{bmatrix} w_1 \\ w_2 \end{bmatrix} \quad [1.31]$$

in equations [1.10] and [1.11] and introduce the notations:

$$\begin{bmatrix} \widetilde{G_1} \\ \widetilde{G_2} \end{bmatrix} = \begin{bmatrix} W_1 \\ W_2 \end{bmatrix} G; \; \begin{bmatrix} \widetilde{T_1} \\ \widetilde{T_2} \end{bmatrix} = \begin{bmatrix} W_1 \\ W_2 \end{bmatrix} T. \quad [1.32]$$

1 Because $\begin{bmatrix} U_1 & U_3 \\ U_2 & U_4 \end{bmatrix}$ is a unitary, it can be shown that $U_4^* = U_4 - U_2 U_1^{-1} U_3$ and $U_3^* + U_4^* U_2 U_1^{-1} = 0$

Equations [1.10] and [1.11] then become:

$$\begin{cases} \dot{w_1} = \widetilde{F_{11}}w_1 + \widetilde{F_{12}}w_2 + \widetilde{G_1}(y - Du) + \widetilde{T_1}Bu \text{ (a)} \\ \dot{w_2} = \widetilde{F_{22}}w_2 + \widetilde{G_2}(y - Du) + \widetilde{T_2}Bu \text{ (b)} \end{cases} \quad [1.33]$$

and

$$\begin{cases} \widetilde{T_1}A - \widetilde{F_{11}}\widetilde{T_1} - \widetilde{F_{12}}\widetilde{T_2} = \widetilde{G_1}C \text{ (a)} \\ \widetilde{T_2}A - \widetilde{F_{22}}\widetilde{T_2} = \widetilde{G_2}C. \text{ (b)} \end{cases} \quad [1.34]$$

Now, we will assume that the Schur decomposition has been performed in such a way that $\widetilde{T_2} = W_2 T$ is non-singular (in proposition 1.3, a necessary condition for T to be full column rank is given) and we perform the second change of variable:

$$w_2 = \widetilde{T_2}\widehat{x}. \quad [1.35]$$

From equations [1.33b] and [1.34b], we can derive:

$$\dot{\widehat{x}} = A\widehat{x} + Bu + \widetilde{T_2}^{-1}\widetilde{G_2}(y - C\widehat{x} - Du). \quad [1.36]$$

Now using [1.34a] and [1.35] to substitute $\widetilde{F_{12}w_2}$ into equation [1.33a], we get:

$$\dot{w_1} = \widetilde{F_{11}}(w_1 - \widetilde{T_1}\widehat{x}) + \widetilde{G_1}(y - C\widehat{x} - Du) + \widetilde{T_1}(A\widehat{x} + Bu). \quad [1.37]$$

Pre-multiplying equation [1.36] by $\widetilde{T_1}$, subtracting it from equation [1.37] and using the last change of variable:

$$w_1 - \widetilde{T_1}\widehat{x} = x_Q, \quad [1.38]$$

we obtain:

$$\dot{x_Q} = \widetilde{F_{11}}x_Q + (\widetilde{G_1} - \widetilde{T_1}\widetilde{T_2}^{-1}\widetilde{G_2})(y - C\widehat{x} - Du). \quad [1.39]$$

From [1.9], [1.31], [1.35] and [1.38], we can easily derive the global linear transformation between the controller original state x_K and the new states $\widehat{x}$ and x_Q:

$$x_K = \widehat{z} = [V_1 \quad T]\begin{bmatrix} x_Q \\ \widehat{x} \end{bmatrix}. \quad [1.40]$$

Then, the controller output equation [1.7b] can be expressed as:

$$u = C_K T\widehat{x} + C_K V_1 x_Q + D_K y \qquad [1.41]$$

or

$$u = J_m \left[(C_K T + D_K C)\widehat{x} + C_K V_1 x_Q + D_K(y - C\widehat{x} - Du)\right]. \qquad [1.42]$$

The identification of the set of equations [1.36], [1.39] and [1.42] with equation [1.3] provides all the parameters for the observer-based controller structure shown in Figure 1.1:

$$K_f = \widetilde{T_2}^{-1}\widetilde{G_2} = (W_2 T)^{-1} W_2 G. \qquad [1.43]$$

$$K_c = -J_m(C_K T + D_K C). \qquad [1.44]$$

$$A_Q = \widetilde{F_{11}} = W_1 F V_1. \qquad [1.45]$$

$$B_Q = \widetilde{G_1} - \widetilde{T_1}\widetilde{T_2}^{-1}\widetilde{G_2} = W_1[I_{n_K \times n_K} - T(W_2 T)^{-1} W_2]G. \qquad [1.46]$$

$$C_Q = J_m C_K V_1. \qquad [1.47]$$

$$D_Q = J_m D_K. \qquad [1.48]$$

REMARK 1.1.– If $\boxed{n_K = n}$, then T is square and decomposition [1.29] of F is such that $V_2 = I_{n\times n}$ and V_1 is empty. Then equations [1.43]–[1.48] become:

$$K_f = T^{-1}G = (T^{-1}B_K - BD_K)J_p. \qquad [1.49]$$

$$K_c = -J_m(C_K T + D_K C). \qquad [1.50]$$

$$Q(s) = D_Q = J_m D_K. \qquad [1.51]$$

This result then specializes to those of [BEN 85].

1.5. Discussion

There is a combinatory of solutions according to the choice of the partition of the closed-loop eigenvalues: first, in the computation of matrix T, and second in the decomposition of matrix F. Hereafter, some rules are proposed to reduce the number of admissible choices.

PROPOSITION 1.1.– *The n eigenvalues chosen for the computation of the solution T of the Riccati equation* [1.18] *are the n state-feedback eigenvalues of the equivalent observer-based controller, i.e. spec$(A - BK_c)$.*

PROOF: From [1.20], [1.21] and [1.22], we have:

$$\begin{bmatrix} A + BJ_mD_KC & BJ_mC_K \\ B_KJ_pC & A_K + B_KDJ_mC_K \end{bmatrix} \begin{bmatrix} I_{n\times n} \\ T \end{bmatrix} = \begin{bmatrix} I_{n\times n} \\ T \end{bmatrix} U_1 \Lambda U_1^{-1}, \qquad [1.52]$$

the first row of this matrix equality reads:

$$A + BJ_m(D_KC + C_KT) = U_1 \Lambda U_1^{-1}, \qquad [1.53]$$

and using [1.44], we have:

$$A - BK_c = U_1 \Lambda U_1^{-1}. \qquad [1.54]$$

So, the eigenvalues of Λ are the eigenvalues of $A - BK_c$.

As a consequence, the n_K remaining eigenvalues are the Luenberger observer eigenvalues (i.e. spec(F), see also equation [1.17]), which are shared, in [1.29], between the $n_K - n$ Youla parameter eigenvalues (i.e. spec(A_Q)) and the n state-estimator eigenvalues (i.e. spec$(A - K_fC)$).

From the set of equations ([1.20], [1.21] and [1.26]), it becomes:

$$\begin{bmatrix} A + BJ_mD_KC & BJ_mC_K \\ B_KJ_pC & A_K + B_KDJ_mC_K \end{bmatrix} \begin{bmatrix} U_1 \\ U_2 \end{bmatrix} = \begin{bmatrix} U_1 \\ U_2 \end{bmatrix} \Lambda, \qquad [1.55]$$

and a necessary condition on the choice of the subspace $\mathcal{S}$ is given for the existence of a solution T (i.e. for U_1 to be invertible).

PROPOSITION 1.2.– *Consider U_1 and U_2 associated with an n-dimensional invariant subspace $\mathcal{S}$ of A_{cl}. If there is an uncontrollable plant eigenvalue which is not in spec$(A_{cl}|S)$ then U_1 is singular. In other words,*

$$if\ \exists\ \lambda \notin spec(\Lambda)\ /\ \lambda\ is\ (A, B)\ uncontrollable,\ then\ U_1\ is\ singular \qquad [1.56]$$

PROOF– Consider the (A, B)-pair and let λ denote an uncontrollable eigenvalue with associated left-eigenvector u. That is,

$$\mathrm{u}^{\mathrm{T}}[A - \lambda I \mid B] = 0, \qquad [1.57]$$

then premultiplying [1.55] by $[\mathrm{u}^{\mathrm{T}}\ 0]$, we get:

$$\mathrm{u}^{\mathrm{T}}[(A + BJ_m D_K C)U_1 + BJ_m C_K U_2] = \mathrm{u}^{\mathrm{T}} U_1 \Lambda. \qquad [1.58]$$

From [1.57] and [1.58] it follows that:

$$\mathrm{u}^{\mathrm{T}} U_1 (\Lambda - \lambda I) = 0. \qquad [1.59]$$

So, if $\lambda \notin \mathrm{spec}(\Lambda)$ then $\mathrm{u}^{\mathrm{T}} U_1 = 0$, that is U_1 is singular.

A dual necessary condition for T to be full column rank (i.e. for U_2 to be full column-rank) can be stated as follows.

PROPOSITION 1.3.– *Consider U_1 and U_2 associated with an n-dimensional invariant subspace $\mathcal{S}$ of A_{cl}. If there is an unobservable plant eigenvalue in spec$(A_{cl}|S)$, then U_2 is column rank deficient. In other words,*

$$\begin{array}{l} if\ \exists\ \lambda\ \in\ spec(\Lambda)\ /\ \lambda\ is\ (A, C)\ unobservable, \\ \qquad then\ U_2\ is\ column\ rank\ deficient. \end{array} \qquad [1.60]$$

PROOF.– Omitted for brevity, see proposition 1.2.

Propositions 1.2 and 1.3 are quite useful when an observer-based realization for H_∞ or μ controllers must be computed from the standard problem augmented with input and output frequency weights (see [ALA 99] for more details).

REMARK 1.2.– Among all the admissible choices, the only restriction which can reduce the set of solutions is that complex auto-conjugate pairs of eigenvalues cannot be separated if a state-space representation with real coefficients is required. Note that such a choice is not always possible. For instance, consider the plant $G_0(s) = 1/s$ and the controller $K_0(s) = 2/(s + 2)$, then the computation of the state-feedback-state-estimator form leads to $Q = 0$, $K_c = 1 + i$ (or $1 - i$) and $K_f = 1 - i$ (respectively $1 + i$). Although the gains K_c and K_f are complex, the transfer function of the controller has real coefficients. It can easily be shown that, for a minimal plant (without uncontrollable or unobservable eigenvalues):

– if n (plant order) is even, then a real solution always exists;

– if n is odd, then a real solution T exits if the number of real eigenvalues in spec(A_{cl}) is at least equal to one and a real parametrization (K_c, K_f, $Q(s)$) exits (in the case, $n_K > n$) if the number of real eigenvalues in spec(A_{cl}) is at least equal to two.

For non-minimal plant, the rule is a little bit more complex. Let $n_{\overline{C}}$ and $n_{\overline{O}}$ be the numbers of plant uncontrollable and unobservable eigenvalues, respectively (the controller is assumed to be minimal). Then:

– if $n - n_{\overline{C}}$ is even, then a real solution T always exists;

– if $n - n_{\overline{C}}$ is odd, then a real solution T exits if the number of real and *controllable* eigenvalues in spec(A_{cl}) is at least equal to one,

– in addition, a real parametrization (K_c, K_f, $Q(s)$) exits (in the case $n_K > n$), if the number of real, *observable* and *controllable* eigenvalues in spec(A_{cl}) is at least equal to rem$(n - n_{\overline{C}}, 2)$ + rem$(n - n_{\overline{O}}, 2)$ + rem$(n_K - n, 2)$, where rem(X, Y) is the remainder after the division of X by Y.

The following selection rules have also proved useful in practical applications of the method (see also section 1.12):

– assign the fastest eigenvalues to spec(A_Q) in such a way that the Youla parameter acts as a direct feedthrough in the controller;

– assign to spec$(A - BK_c)$ the n closed-loop eigenvalues which are the "nearest" from the n plant eigenvalues in order to respect the dynamic behavior of the physical plant and reduce the state-feedback gains;

– assign fast closed-loop eigenvalues to spec$(A - K_f C)$ to have an efficient state-estimator.

1.6. In brief

The procedure to compute the observer-based realization and the dynamic Youla parameter for a given n_Kth-order controller associated with an nth-order plant ($n_K \geq n$) can be summarized as follows:

– compute the closed-loop matrix A_{cl} (equations [1.20] and [1.8]) and split up the $n + n_K$ eigenvalues of A_{cl} into three auto-conjugate sets:

- n eigenvalues to be assigned to state-feedback dynamics spec$(A - BK_c)$,

- $n_K - n$ eigenvalues to be assigned to the Youla parameter dynamics spec(A_Q),

- n eigenvalues to be assigned to state-estimator dynamics spec$(A - K_f C)$;

– compute a Schur or a diagonal decomposition of A_{cl} (equation [1.26]) such that the eigenvalues are ordered on the diagonal according to the previous choice; that is, $\mathrm{spec}(\Lambda) = \mathrm{spec}(A - BKc)$ and $\mathrm{spec}(\Lambda^F) = \mathrm{spec}(A_Q) \cup \mathrm{spec}(A - KfC)$;

– compute T, F and G with equations [1.23], [1.14], [1.13],

– compute V, V_1, V_2, W_1, W_2 with equations [1.28], [1.30];

– compute the sought parameters K_c, K_f, A_Q, B_Q, C_Q and D_Q using [1.43]–[1.48] and [1.8].

The reader will find an interactive Matlab® function named `cor2obr` (in the toolbox `bib_obr`) to compute the observer-based realization for a given controller and a given plant at: http://personnel.isae.fr/daniel-alazard/matlab-packages. A demo file called `demo_obr.m` illustrates how to use this function to compute an observer-based realization of the 16th-order controller designed in the demo #3 of the `Mu-Analysis` and `Synthesis Toolbox`.

The `help` of the function `cor2obr.m` is given in section A4.2, Appendix 4. Section A4.3 gives the `help` of the reciprocal function `obr2cor`, that is: `Kss=obr2cor(G0,Kc,Kf,Q)` allows the state-space realization of a controller (equation [1.3]) to be computed from its observer-based parametrization (`Kc,Kf,Q`) associated with a given plant `G0`.

1.7. Reduced-order controllers case

In the case $n_K < n$ (i.e. $dim(z) < dim(x)$), the observer-based structure shown in Figure 1.1 is no longer valid. But an interesting alternative can be derived using a reduced-order estimator.

It is interesting to point out the case where $[T^{\mathrm{T}}\ C^{\mathrm{T}}]$ is a rank n matrix (i.e. $p + n_K \geq n$). Then, a reduced observer-based realization involving an estimate $\widehat{x}$ of the plant state x can be obtained by a linear combination of the controller state $\widehat{z}$ and the plant output y (see [LUE 71]):

$$\widehat{x} = H_1\widehat{z} + H_2(y - Du), \qquad [1.61]$$

with the constraint:

$$H_1T + H_2C = I_n. \qquad [1.62]$$

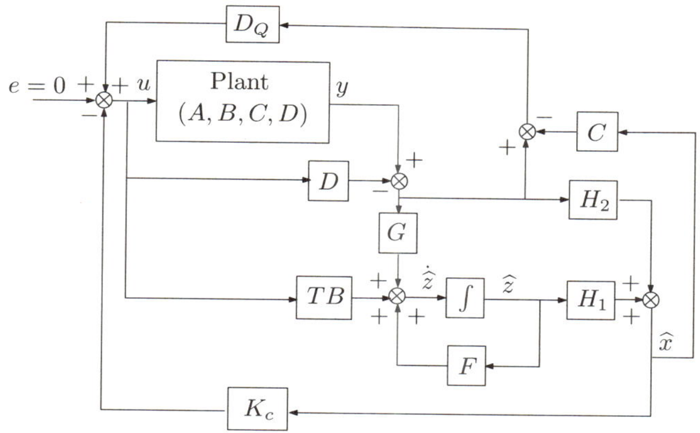

Figure 1.7. *Reduced-order observer-based realization of the controller*

Then, the separation principle still holds and a Youla parametrization (with a static parameter D_Q) built on such a reduced-order estimator reads:

$$\begin{cases} \dot{\widehat{z}} = F\widehat{z} + G(y - Du) + TBu & \text{(a)} \\ \widehat{x} = H_1\widehat{z} + H_2(y - Du) & \text{(b)} \\ u = -K_c\widehat{x} + D_Q(y - C\widehat{x} - Du). & \text{(c)} \end{cases} \quad [1.63]$$

$$\begin{cases} TA - FT = GC \\ H_1T + H_2C = I_n. \end{cases} \quad [1.64]$$

The structure of the controller is then depicted in Figure 1.7. The state-space representation of the controller $K_0(s)$ between $y - Du$ and u (the plant direct feedthrough matrix D acts in negative feedback on the controller) also reads:

$$\left[\begin{array}{c} \dot{\widehat{z}} \\ \hline u \end{array}\right] = \left[\begin{array}{c|c} F - TB(K_c + D_QC)H_1 & G + TB(D_Q - (K_c + D_QC)H_2) \\ \hline -(K_c + D_QC)H_1 & D_Q - (K_c + D_QC)H_2 \end{array}\right] \left[\begin{array}{c} \widehat{z} \\ \hline y - Du \end{array}\right]. \quad [1.65]$$

As previously described, it can easily be shown that the closed-loop eigenvalues, with a controller defined by equations [1.63] and [1.64], are distributed between the n state-feedback eigenvalues (spec$(A - BK_c)$) and the n_K estimator eigenvalues (spec(F)). Equations [1.18], [1.14] and [1.13] which respectively provide T, F and G are still valid. The problem is therefore reduced to compute K_c, H_1, H_2 and D_Q such that (from the identification of [1.63b] and [1.63c] with [1.12b] rewritten as $u = J_mC_k\widehat{x} + J_mD_K(y - Du)$):

$$\begin{cases} J_mC_K = -(K_c + D_QC)H_1 & (a) \\ J_mD_K = -(K_c + D_QC)H_2 + D_Q & (b) \\ H_1T + H_2C = I_n. & (c) \end{cases} \quad [1.66]$$

It is easily deduced that:

$$K_c = -J_m(C_K T + D_K C). \quad [1.67]$$

This is equation [1.44], established in the augmented-order controller case.

To compute H_1, H_2 and D_Q, the following situations can be considered:

– if $\begin{bmatrix} T \\ C \end{bmatrix}^{-1}$ exists (which implies that $n_K + p = n$) then:

$$\boxed{[\underbrace{H_1}_{n_K} \quad \underbrace{H_2}_{p}] = \begin{bmatrix} T \\ C \end{bmatrix}^{-1}} \quad [1.68]$$

and

$$\begin{bmatrix} T \\ C \end{bmatrix} [H_1 \quad H_2] = \begin{bmatrix} TH_1 & TH_2 \\ CH_1 & CH_2 \end{bmatrix} = \begin{bmatrix} I_{n_K} & 0 \\ 0 & I_p \end{bmatrix}. \quad [1.69]$$

Hence, relationships [1.66] hold true for any D_Q and we can choose:

$$\boxed{D_Q = 0} \quad [1.70]$$

without loss of generality. This case is illustrated at the end of section 1.8.1.

– if $n_K > n - p$, then there are several solutions (H_1 and H_2) satisfying [1.66], we can choose, for example, the least norm solution (in order to reduce the control gains) using the pseudo-inverse of matrix $[T^{\mathrm{T}}\ C^{\mathrm{T}}]$:

$$\boxed{\begin{cases} H_1 = [T^{\mathrm{T}}T + C^{\mathrm{T}}C]^{-1}T^{\mathrm{T}} \\ H_2 = [T^{\mathrm{T}}T + C^{\mathrm{T}}C]^{-1}C^{\mathrm{T}} \end{cases}}. \quad [1.71]$$

Then, from [1.66]:

$$\boxed{D_Q = (J_m D_K + K_C H_2)(I_p - CH_2)^{-1}) \ .}$$

If $n_K < n - p$, it can only be stated that, in open-loop, the controller state $\widehat{z}$ is an estimate of the linear combination T of the plant state x, that is the estimation error $\varepsilon_z = Tx - \widehat{z}$ tends to 0 with the following dynamics:

$$\dot{\varepsilon_z} = (A_K + (B_K D - TB)J_m C_K)\varepsilon_z \quad [1.72]$$

In this case ($n_K < n - p$), the only way round consists of performing a reduction of the plant until the previous situation is applicable. The controller is then interpreted as an observer-based controller associated with the reduced model of the plant.

In section 1.8, the interest of observer-based realizations of a given controller is highlighted through three examples: plant state monitoring, controller switching and smooth gain scheduling on an academic second-order model of a launcher.

1.8. Illustrations

Let us consider again the model of a launcher between the angle of attack α and the thruster deflection δ (see section 1.3):

$$G_0(s) = \frac{1}{s^2 - 1}$$

associated with the state-space realization:

$$\begin{bmatrix} \dot{\alpha} \\ \ddot{\alpha} \\ \hline \alpha \end{bmatrix} = \left[\begin{array}{cc|c} 0 & 1 & 0 \\ 1 & 0 & 1 \\ \hline 1 & 0 & 0 \end{array}\right] \begin{bmatrix} \alpha \\ \dot{\alpha} \\ \hline \delta \end{bmatrix}. \qquad [1.73]$$

Let us consider the following stabilizing controller (positive feedback):

$$K_1(s) = -\frac{s^2 + 27s + 26}{s^2 + 7s + 18}.$$

A state-space realization (companion form[2]) of this controller reads:

$$\begin{bmatrix} \dot{x_1} \\ \dot{x_2} \\ \hline \delta \end{bmatrix} = \left[\begin{array}{cc|c} 0 & -18 & 1 \\ 1 & -7 & 0 \\ \hline -20 & 132 & -1 \end{array}\right] \begin{bmatrix} x_1 \\ x_2 \\ \hline \alpha \end{bmatrix}. \qquad [1.74]$$

In this example, the closed-loop dynamics reveals multiple eigenvalues:

$$\text{spec}(A_{cl}) = \{-2, \quad -2, \quad -2, \quad -1\}.$$

Then, there exists two admissible choices to solve in T the Riccati equation [1.18]. The choice $\text{spec}(A - BK_c) = \{-1, \quad -2\}$ and the application of the procedure provide the following parametrization:

$$K_c = [3 \quad 3]; \; K_f = [4 \quad 5]^{\mathrm{T}}; \; Q = -1.$$

2 Such a vertical companion form can easily be obtained using the Matlab® macro-function `canon`.

Then, the observer-based realization of $K_1(s)$ (equation [1.3]) reads:

$$\begin{bmatrix} \dot{\hat{\alpha}} \\ \dot{\hat{\dot{\alpha}}} \\ \hline \delta \end{bmatrix} = \left[\begin{array}{cc|c} -4 & 1 & 4 \\ -6 & -3 & 4 \\ \hline -2 & -3 & -1 \end{array}\right] \begin{bmatrix} \hat{\alpha} \\ \hat{\dot{\alpha}} \\ \hline \alpha \end{bmatrix} \qquad [1.75]$$

associated with the estimated state vector $\hat{x} = [\hat{\alpha},\ \hat{\dot{\alpha}}]^{\mathrm{T}}$.

The corresponding Matlab® sequence using functions cor2obr and obr2cor is:

```
G0=ss([0 1;1 0],[0;1],[1 0],0);
cor=tf(-[1 27 26],[1 7 18]);
K1=canon(cor,'companion');
[Kc,Kf,Q]=cor2obr(G0,cor) % other syntax: OBdata=cor2obr(G0,cor),
K1ob=obr2cor(G0,Kc,Kf,Q)  %               K1ob=OBdata.obr
```

A Matlab® demo file demo_section_1_8.m for the following illustrations is also available at: http://personnel.isae.fr/daniel-alazard/matlab-packages.

1.8.1. *Illustration 1: plant state monitoring*

Figures 1.8 and 1.9 illustrate the closed-loop state responses (launcher and controller states) to initial conditions on launcher states ($\alpha(t = 0) = 1$ rad and $\dot{\alpha}(t = 0) = -1$ rad/s). Figure 1.8 is obtained when the first controller realization (equation [1.74]) is used, while Figure 1.9 is obtained with the observer-based realization (equation [1.75]). For both simulations the launcher state responses are the same because the initial conditions are the same and the input–output behavior of the controller is independent of its realization. But, we can see in Figure 1.8 that there is no straightforward relation between controller states and launcher states (α and $\dot{\alpha}$), while Figure 1.9 highlights that (after the transient response of the state-estimator) the controller states of the observer-based realization are a good estimate of launcher states and can be used to monitor launcher states for offline or online analysis (for failure diagnosis purposes, for instance). As the plant state are meaningful variables (α (rad) and $\dot{\alpha}$ (rad/s)), we can also conclude that the state-feedback gain K_c has a physical unit: $K_c = [3\,\mathrm{rad/rad}\quad 3\,\mathrm{s}]$, while the dimension of the various components of realization [1.74] is not defined.

1.8.1.1. *Reduced-order controller case*

Let us consider the first-order controller $K_0(s) = -23s - 32/s + 12$ introduced in section 1.3. The following Matlab® sequence illustrates the result of section 1.7: that is, the way to monitor launcher states from the measurement α and the single controller state x_K (defined in equation[1.25]). This sequence follows the sequence presented in page 11 (section 1.3) and provides responses presented in Figure 1.10:

responses of the angle of attack α and its estimate $\widehat{\alpha}$ are the same because α is measured but the response of the estimated angle of attack rate $\widehat{\dot{\alpha}}$ converges to the launcher state $\dot{\alpha}$ with the estimation dynamics (here $F = -10$ rad/s), that is in less than half a second.

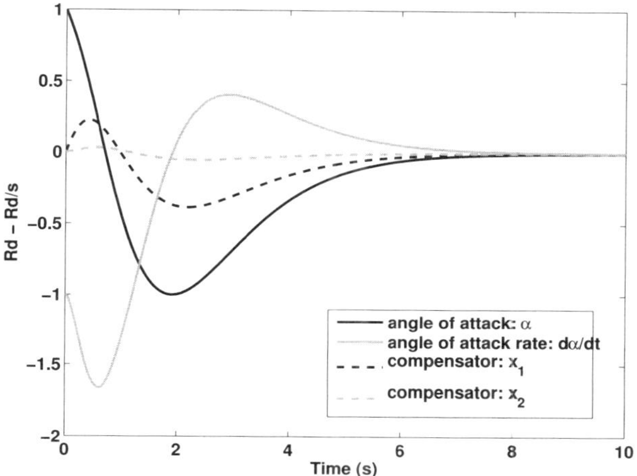

Figure 1.8. *Responses to initial conditions on launcher states – companion realization of $K_1(s)$ (equation [1.74])*

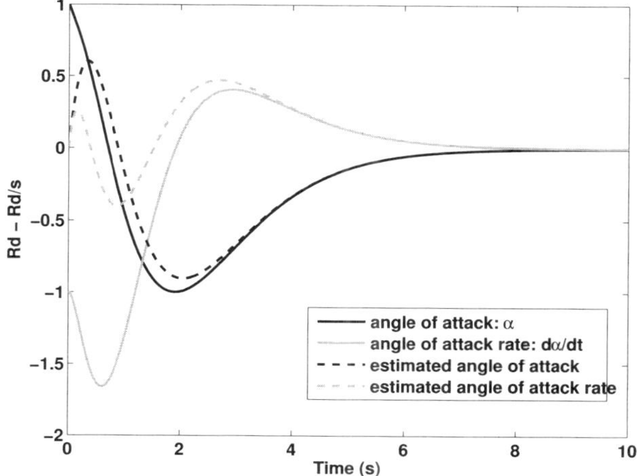

Figure 1.9. *Responses to initial conditions on launcher states – observer-based realization of $K_1(s)$ (equation [1.75])*

```
H1H2=inv([T;G0.c]);
H1=H1H2(:,1);
H2=H1H2(:,2);
xhat=x(:,3)*H1'+y*H2';
figure plot(t,x(:,1),'k-','LineWidth',1)
hold on plot(t,x(:,2),'g-','LineWidth',2)
plot(t,xhat(:,1),'k—','LineWidth',2)
plot(t,xhat(:,2),'g—','LineWidth',2)
xlabel('Time (s)'); ylabel('Rd - Rd/s');
```

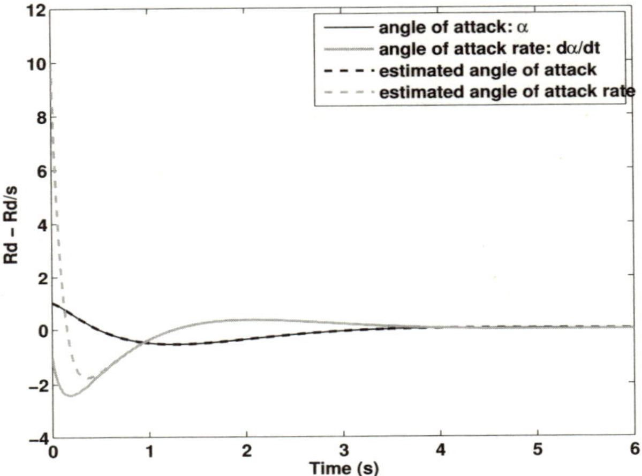

Figure 1.10. *Responses to initial conditions on launcher states with first-order controller* $K_0(s)$

1.8.2. *Illustration 2: controller switching*

Let us consider a second stabilizing controller:

$$K_2(s) = -\frac{1667s + 2753}{s^2 + 27s + 353}$$

and let us assume that the control law must switch from controller K_1 to controller K_2 at time $t = 5$ s. This new controller increases closed-loop dynamic performances required, for instance, during the final flight phase. Indeed, the closed-loop dynamics is now:

$$\text{spec}(A_{cl}) = \{-3, \quad -4, \quad -10 + 10\,i, \quad -10 - 10\,i\}.$$

Note that the structure of this new controller K_2 is quite different from the previous structure (the direct feed through term is null in K_2). An observer-based parametrization for $K_2(s)$ reads[3]:

$$K_c = [13 \quad 7];\; K_f = [20 \quad 201]^{\mathrm{T}};\; Q = 0.$$

The companion realization of this controller provided by Matlab® macro-function canon reads:

$$\begin{bmatrix} \dot{x_1} \\ \dot{x_2} \\ \hline \delta \end{bmatrix} = \left[\begin{array}{cc|c} 0 & -353 & 1 \\ 1 & -27 & 0 \\ \hline -1667 & 42260 & 0 \end{array}\right] \begin{bmatrix} x_1 \\ x_2 \\ \hline \alpha \end{bmatrix}. \qquad [1.76]$$

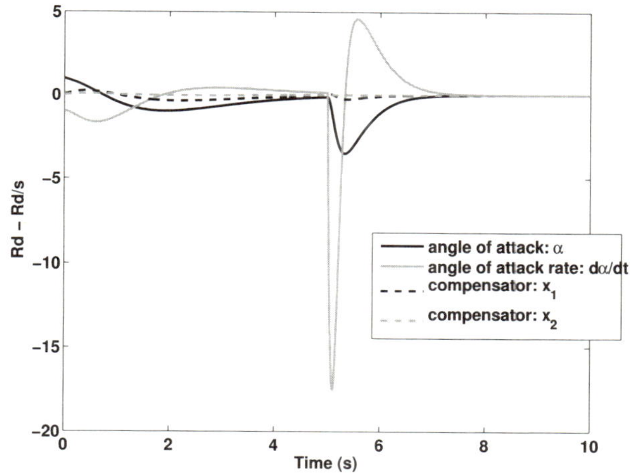

Figure 1.11. *Responses to initial conditions and switch from $K_1(s)$ to $K_2(s)$ at time $t = 5$ s – companion realizations of $K_i(s)$*

The state vector initialization of the second controller K_2 with the value of the state vector of the first controller at the switch time (5 s) can create an undesirable transient response (see Figure 1.11 when vertical companion realizations are used for K_1 and K_2). The meaningful state of the observer-based realizations of both controllers allows us to initialize correctly the second controller and so allows the transient response on the attitude $\alpha(t)$ to be reduced in a significant way (see Figure 1.12). Of course the best improvement can be seen on the response of the control signal δ (thruster

3 This observer-based parametrization assigns the two real closed-loop eigenvalues (i.e. -3 and -4) to the state-feedback dynamics.

deflection) depicted in Figure 1.13 (for companion form switching) and Figure 1.14 (for observer-based form switching): the transient peak value at the switching time is considerably reduced when observer-based realizations are used.

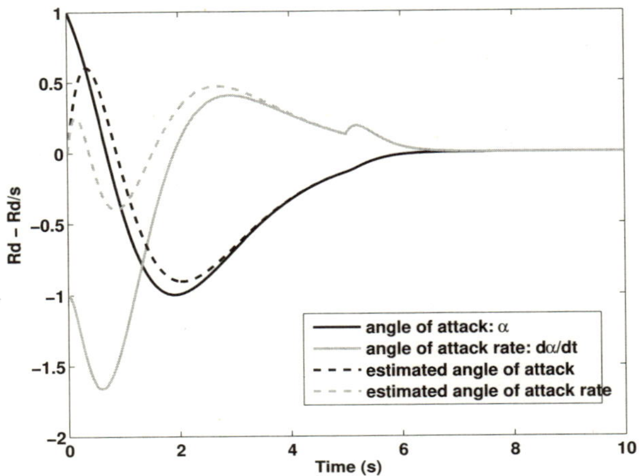

Figure 1.12. *Responses to initial conditions and switch from $K_1(s)$ to $K_2(s)$ at time $t = 5$ s – observer-based realizations of $K_i(s)$*

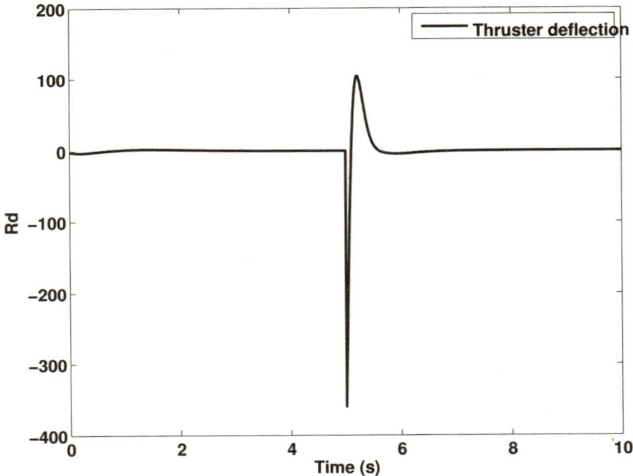

Figure 1.13. *Responses of the control signal δ to initial conditions and switch from $K_1(s)$ to $K_2(s)$ at time $t = 5$ s – companion realizations of $K_i(s)$*

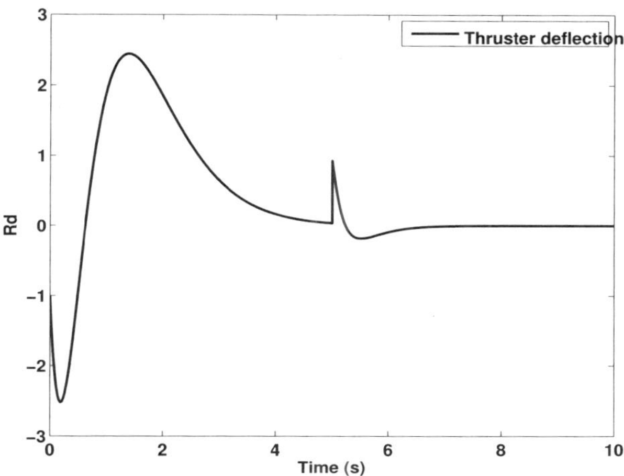

Figure 1.14. *Responses of the control signal δ to initial conditions and switch from $K_1(s)$ to $K_2(s)$ at time $t = 5$ s – observer-based realizations of $K_i(s)$*

1.8.3. *Illustration 3: smooth gain scheduling*

Now, let us assume that we wish to interpolate the controller from K_1 to K_2 over 5 s (starting from $t = 0$). The linear interpolation of the four state-space matrices of companion realizations provides a non-stationary controller $K(s,t)$ whose frequency response w.r.t. time t is depicted in Figure 1.15. We can note that this response is non-monotonous at low frequency and we can also easily check that the controller $K(s,t)$ does not stabilize the plant $G_0(s)$ for all t between 0.1 and 4.6. For instance, at $t = 2.5$ s the current controller $K(s, t = 2.5\,\text{s})$ is defined by the companion realization (i.e. mean values of state-space matrices defined in [1.74] and [1.76]):

$$\begin{bmatrix} \dot{x_1} \\ \dot{x_2} \\ \hline \delta \end{bmatrix} = \left[\begin{array}{cc|c} 0 & -185.5 & 1 \\ 1 & -17 & 0 \\ \hline -843.5 & 21190 & -0.5 \end{array}\right] \begin{bmatrix} x_1 \\ x_2 \\ \hline \alpha \end{bmatrix}. \qquad [1.77]$$

That is: $K(s, t = 2.5\,\text{s}) = -0.5(s + 1712)(s - 7.9)/s^2 + 17\text{s} + 185.5$, which does not stabilize $G_0(s)$.

The interpolation of the four state-space matrices of observer-based realizations of K_1 and K_2 provides a smoother interpolation (see Figure 1.16). We can also check that this new interpolated controller stabilizes $G_0(s)$ for all time $t \in [0,\ 5\,\text{s}]$.

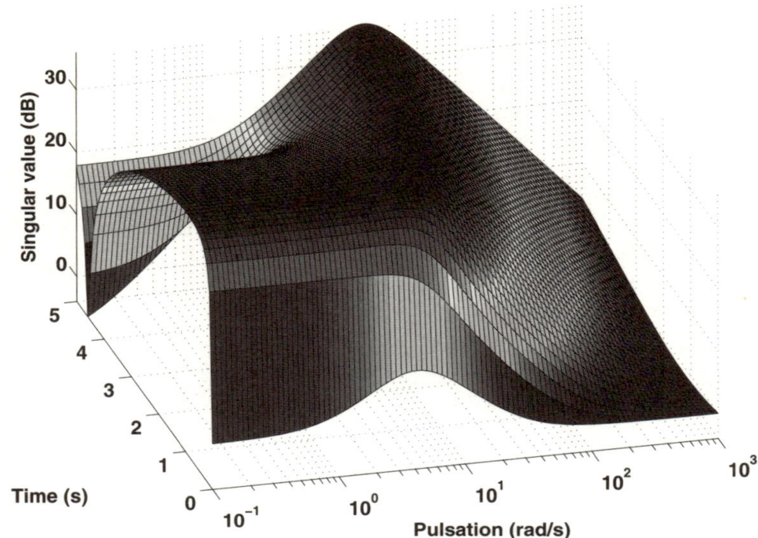

Figure 1.15. *$K(s,t)$: singular value w.r.t time*

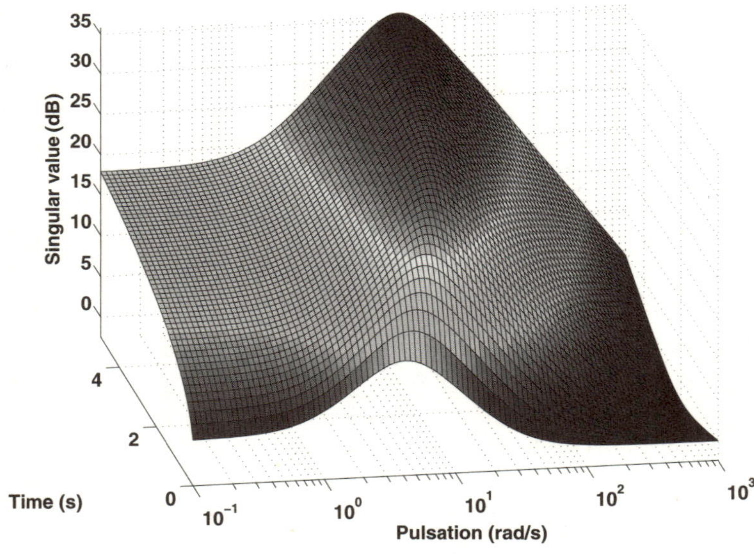

Figure 1.16. *$K_{observer-based}(s,t)$: singular value w.r.t time*

1.9. Reference inputs in observer-based realizations

1.9.1. *General results*

Let us consider one more time the observer-based control depicted in Figure 1.1. A_Q, B_Q, C_Q and D_Q are the four of the state-space matrices of $Q(s)$ (order n_Q) associated with the state variable x_Q. Let us assume that the signal $e(t)$ is no longer null but is defined by:

$$e(t) = Hr(t) \tag{1.78}$$

where $r(t)$ is the reference input signal with l components and $H_{m\times l}$ is a static feed forward matrix. The objective is now to compute the state-space realization of the two degrees of freedom (2 dof) controller between the controller inputs $[y^{\mathrm{T}} \quad r^{\mathrm{T}}]^{\mathrm{T}}$ and the controller output u.

From Figure 1.1, we can easily derive:

$$u = -K_c\widehat{x} + C_Q x_Q + D_Q y - D_Q C\widehat{x} - D_Q D u + e.$$

Let J_m^+ be defined[4] by $\boxed{J_m^+ = (I_m + D_Q D)^{-1}}$, then the output equation of the 2 dof controller reads:

$$u = C_{obr}\begin{bmatrix}\widehat{x}\\ x_Q\end{bmatrix} + D_{obr}\begin{bmatrix}y\\ r\end{bmatrix}$$

with

$$\boxed{\begin{aligned} C_{obr} &= -J_m^+[K_c + D_Q C \quad -C_Q]. && [1.79]\\ D_{obr} &= J_m^+[D_Q \quad H]. && [1.80]\end{aligned}}$$

From Figure 1.1, we can also derive:

$$\dot{\widehat{x}} = (A - K_f C)\widehat{x} + K_f y + (B - K_f D)u. \tag{1.81}$$

$$\dot{x_Q} = -B_Q C\widehat{x} + A_Q x_Q + B_Q y - B_Q D u. \tag{1.82}$$

4 This notation is used to refer to equation [1.8] previously defined. Indeed, if D_Q is the result of equation [1.48] then it can easily be shown that $J_m^+ = J_m^{-1} = I_m - D_K D$.

Then the observer-based realization of the 2 dof controller reads:

$$\begin{bmatrix} \dot{\widehat{x}} \\ \dot{x_Q} \\ \hline u \end{bmatrix} = \left[\begin{array}{c|c} A_{obr} & B_{obr} \\ \hline C_{obr} & D_{obr} \end{array}\right] \begin{bmatrix} \widehat{x} \\ x_Q \\ \hline y \\ r \end{bmatrix} \qquad [1.83]$$

$$A_{obr} = \begin{bmatrix} A - K_f C & 0 \\ -B_Q C & A_Q \end{bmatrix} + \begin{bmatrix} B - K_f D \\ -B_Q D \end{bmatrix} C_{obr}. \qquad [1.84]$$

with $$B_{obr} = \begin{bmatrix} K_f & 0 \\ B_Q & 0 \end{bmatrix} + \begin{bmatrix} B - K_f D \\ -B_Q D \end{bmatrix} D_{obr}. \qquad [1.85]$$

In a more general context, it is also possible to take into account an additional input matrix $M = \begin{bmatrix} M_{1_{n\times m}} \\ M_{2_{n_Q\times m}} \end{bmatrix}$ between $e = Hr$ and $\begin{bmatrix} \dot{\widehat{x}} \\ \dot{x_Q} \end{bmatrix}$, that is the input matrix B_{obr} becomes:

$$B_{obr} = \begin{bmatrix} K_f & M_1 H \\ B_Q & M_2 H \end{bmatrix} + \begin{bmatrix} B - K_f D \\ -B_Q D \end{bmatrix} D_{obr}.$$

In Franklin and Johnson [FRA 81], it is shown that the matrix M allows us to assign $n + n_Q$ transmission zeros in the closed-loop transfer between the reference input r and the plant output y. The case $M = 0$ corresponds to the particular case where these $n + n_Q$ transmission zeros are assigned to the state estimation dynamics (spec$(A - K_f C)$) and the Youla parameter dynamics spec(A_Q). These poles/zeros cancellations express that these dynamics are uncontrollable from e (see section 1.2). Therefore, considering a controller $K_0(s)$ previously designed to meet some disturbance rejection specifications or some dynamics (eigenvalues) assignment specifications, it is possible to shape the response to the reference input (without any alteration of previous loop properties) using an observer-based realization of this controller where the state-feedback dynamics (spec$(A - BK_c)$) has been judiciously chosen. Indeed, the response of the plant to reference input depends only on this state-feedback dynamics: if spec$(A - BK_c)$ is fast and correctly damped, this response will also be fast and damped. Of course, such a choice is antagonistic with the choice proposed at the end of section 1.5 where it was proposed to assign fast closed-loop eigenvalues to spec$(A - K_f C)$ to have an efficient state-estimator. These considerations are explained in section 1.9.2. The toolbox `bib_obr` also (see http://personnel.isae.fr/daniel-alazard/matlab-packages) includes a Matlab® function `obr2cor2ddl` to compute the state-space matrices of the two dof controller defined in equation [1.83] (the `help` of this function can be found in section A4.4, Appendix 4).

From a practical point of view, the observer-based realization allows us to take into account an equilibrium state vector x_e and the associated equilibrium input u_e as reference inputs. We have just to choose $e = u_e + K_c x_e$ according to the control structure depicted in Figure 1.17. That is to say, the applied control signal is $u = u_e + K_c(x_e - \widehat{x})$. x_e and u_e can be provided by an outer loop (guidance loop) and must satisfy:

$$\dot{x_e} = 0 = Ax_e + Bu_e.$$

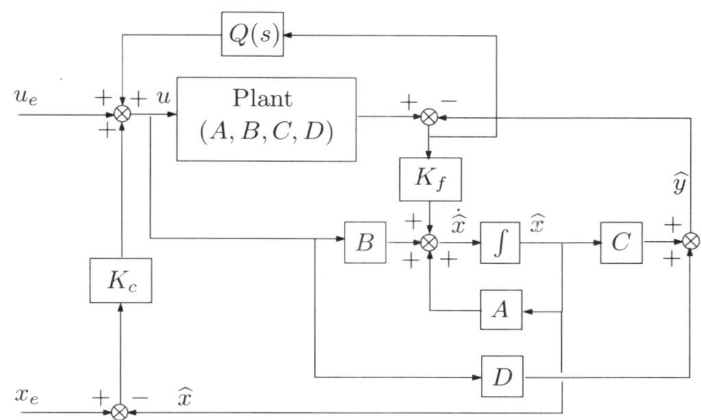

Figure 1.17. *Observer-based controller with equilibrium state x_e and input u_e as reference inputs*

NOTE.– In the Single-Input Single-Output (SISO) case, if the plant exhibits one integrator, that is: A is rank deficient, it is always possible to choose $x_e \in \text{Ker}(A)$ in such a way that $u_e = 0$. Otherwise, A^{-1} exists, then $x_e = -A^{-1}Bu_e$ and $u = (I_m - K_c A^{-1} B)u_e - K_c \widehat{x}$. If $r(t)$ in [1.78] corresponds to u_e, then $H = (I_m - K_c A^{-1} B)$.

Because of the property of the observer-based structure, we can guarantee that, in steady state and if the closed-loop is stable:

$$x(t \to \infty) = \widehat{x}(t \to \infty) = x_e \quad \text{and} \quad u(t \to \infty) = u_e \ .$$

1.9.2. *Illustration*

Let us consider again the model $G_0(s)$ of a launcher between the angle of attack (α) and the thrust deflection (δ) defined by equation [1.73]. Let us consider now a third controller defined by:

$$K_3(s) = -\frac{50s + 70}{s^2 + 10s + 50}. \qquad [1.86]$$

The closed-loop dynamics of $K_3(s)$ in positive feedback with G_0 (see Figure 1.18) is composed of two pairs of auto-conjugate eigenvalues. The first is slow (0.703 rad/s), the second is fast (6.36 rad/s):

Eigenvalue	Damping	Frequency (rad/s)
−4.38e−001 + 5.49e−001i	6.24e−001	7.03e−001
−4.38e−001 − 5.49e−001i	6.24e−001	7.03e−001
−4.56e+000 + 4.44e+000i	7.17e−001	6.36e+000
−4.56e+000 − 4.44e+000i	7.17e−001	6.36e+000

The reference input $r(t)$ is plugged into the loop through the static-feedforward gain H according to Figure 1.18 (positive feedback). In such a classical feedback loop, H is computed to have a unit steady-state gain (DC gain) between r and y, that is $H = 0.4$.

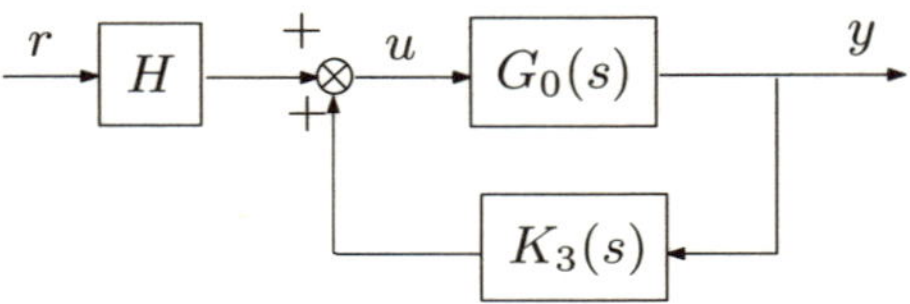

Figure 1.18. *Positive feedback connection of plant $G_0(s)$ and controller $K_3(s)$ with reference input $r(t)$ and static-feedforward H*

The response of $y = \alpha(t)$ to a square signal on $r(t) = \alpha_{ref}(t)$ (i.e. a reference input on the angle of attack) and to initial conditions on launcher states ($\alpha(t = 0) =$ 1 rad and $\dot{\alpha}(t = 0) = -1$ rad/s) is depicted in Figure 1.19. This response exhibits a large settling time and is obviously governed by the slow dynamics.

Considering the observer-based realization depicted in Figure 1.17, the reference equilibrium state is $x_e = [\alpha_{\text{ref}} \quad 0]^{\text{T}}$ and the equilibrium condition $Ax_e + Bu_e = 0$ allows the associated equilibrium input u_e to be computed: $u_e = -\alpha_{\text{ref}}$. The classical feedback loop requires the computation of H (dependent on the controller $K_3(s)$) and the observer-based implementation requires the computation of u_e (independent of the controller) which is quite equivalent. But keeping in mind that such a control law can be the inner loop of an outer guidance loop, the observer-based implementation is very interesting to track more complex reference state trajectories (for instance with a profile on the angle of attack rate $\dot{\alpha_{\text{ref}}}$).

There are two solutions in the computation of the observer-based realization:

– it is possible to assign the slow dynamics (0.703 rad/s) to the state-feedback dynamics. Then the observer-based parametrization reads:

$$K_c = [1.4937 \quad 0.8762]; \quad K_f = [9.1238 \quad 41.5123]^{\text{T}} \text{ and } Q = 0.$$

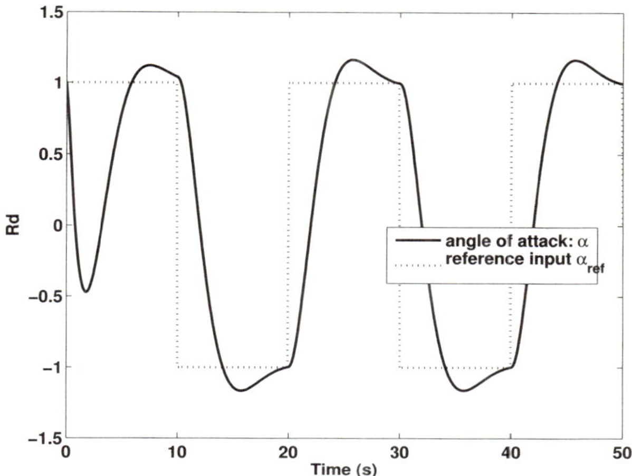

Figure 1.19. *Time-domain responses of classical loop (see Figure 1.18)*

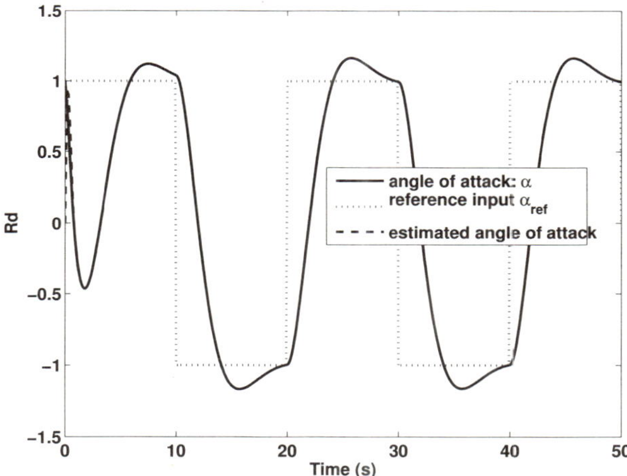

Figure 1.20. *Time-domain responses using observed-based realization (slow dynamics* $(0.703$ *rad/s) is assigned to spec*$(A - BK_c))$

The response for the same reference input profile and the same initial conditions is depicted in Figure 1.20. Performances are very close to the previous response (Figure 1.19). Of course, the estimation of the angle of attack is very good as the state estimation dynamics is fast (see the zoom around $t = 0$ in Figure 1.21: the transient response $\alpha - \widehat{\alpha}$ is very short.).

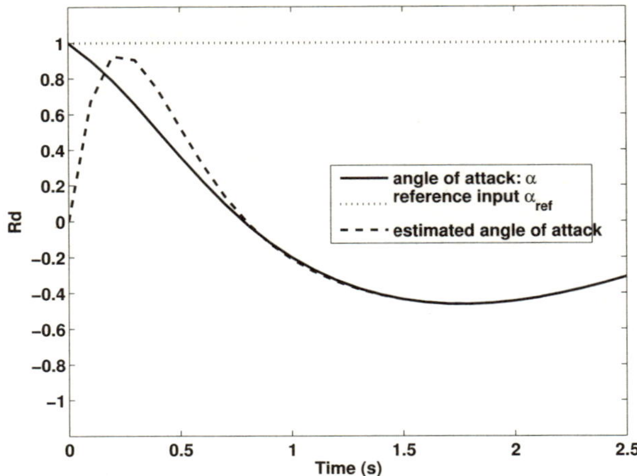

Figure 1.21. *Time-domain responses using observed-based realization (slow dynamics (*0.703 *rad/s) is assigned to spec*$(A - BK_c)$*) – zoom around* $t = 0$

– it is possible to assign the fast dynamics (6.36 rad/s) to the state-feedback dynamics. Then the observer-based parametrization reads:

$$K_c = [41.5123 \quad 9.1238]; \quad K_f = [0.8762 \quad 1.4937]^{\mathrm{T}} \text{ and } Q = 0.$$

The response is then depicted in Figure 1.22: the reference input tracking performance is now very good (once the transient response due to initial estimation error, longer than in the previous case, is finished).

This simple example show that there are some interesting alternatives, based on the observer-based realization of the controller, in the way to plug the reference input in a servo-loop.

1.10. Disturbance monitoring and rejection

1.10.1. *General results*

In section 1.7, the reduced-order observer-based structure was presented. An interesting application consists of increasing the plant model with a model of external disturbances in an augmented model $G_a(s)$. This way and while the order of the augmented model G_a is lower or equal to n_K (controller order) + p (number of measurements), the state of the controller and the measurements can be used to monitor external disturbances. In this section, a low-frequency disturbance $d(t)$ on the input of the nominal plant $G_0(s)$ is considered (see Figure 1.23). Without loss of generality, the following assumption are made:

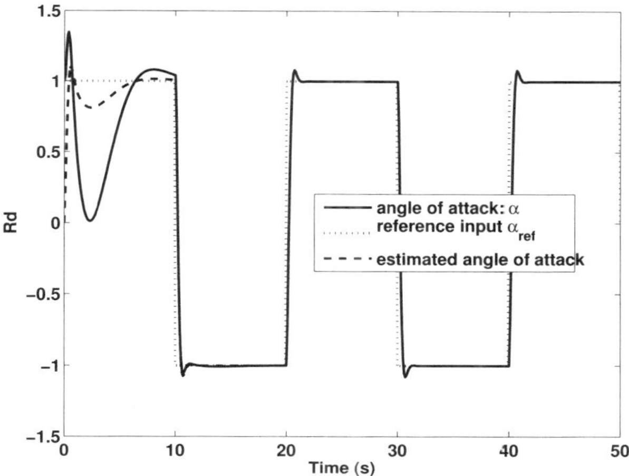

Figure 1.22. *Time-domain responses using observed-based realization (fast dynamics (6.36 rad/s) is assigned to spec$(A - BK_c)$)*

– the direct feed through D of $G_0(s)$ is assumed to be null: $D = 0$ in equation [1.2];

– $G_0(s)$ is assumed to be a single input plant: $m = 1$;

– $n_K + p = n + 1$.

The input disturbance is modeled as the output of a first-order system with a dynamics $\lambda < 0$ and an unknown initial condition $d(0)$:

$$d(t) = d(0)e^{\lambda t} \quad \Rightarrow \quad \dot{d}(t) = \lambda d(t). \qquad [1.87]$$

Thus, the augmented model $G_a(s)$ (order $n + 1$) considered to compute the observer-based realization of the controller reads:

$$\begin{bmatrix} \dot{x} \\ \dot{d} \\ \hline y \end{bmatrix} = \left[\begin{array}{cc|c} A & B & B \\ 0_{n\times 1} & \lambda & 0 \\ \hline C & 0 & 0 \end{array}\right] \begin{bmatrix} x \\ d \\ \hline u \end{bmatrix} = \left[\begin{array}{c|c} A_a & B_a \\ \hline C_a & 0 \end{array}\right] \begin{bmatrix} x \\ d \\ \hline u \end{bmatrix}.$$

It is also assumed that λ is chosen in such a way that the pair (A_a, C_a) is observable.

Applying the procedure described in section 1.7 (more particularly equations [1.18], [1.14], [1.13], [1.67], [1.68] and [1.70]), the observer-based parametrization on the augmented model $G_a(s)$ of the given controller $K_0(s)$ can be determined:

$$F_{(n_K \times n_K)},\ G_{(n_K \times p)},\ T_a = [T_{(n_k \times n)}\ \ T_{d_{(n_k \times 1)}}],\ K_{c_a} = [K_{c_{(1\times n)}}\ \ K_d],$$

$$H_{1_a} = \begin{bmatrix} H_{1_{(n\times n_K)}} \\ H_{1_{d_{(1\times n_K)}}} \end{bmatrix},\ H_{2_a} = \begin{bmatrix} H_{2_{(n\times p)}} \\ H_{2_{d_{(1\times p)}}} \end{bmatrix},\ D_Q = 0_{(1\times p)}$$

such that:

$$T_a A_a - G C_a = F T_a \quad \text{and} \quad [H_{1_a}\ \ H_{2_a}] \begin{bmatrix} T_a \\ C_a \end{bmatrix} = I_{n+1}.$$

The development of these two equations leads to:

$$TA - GC = FT \qquad [1.88]$$

$$TB = (F - \lambda I_{n_K}) T_d \qquad [1.89]$$

$$H_1 T + H_2 C = I_n \qquad [1.90]$$

$$H_{1_d} T + H_{2_d} C = 0_{(1\times n)} \qquad [1.91]$$

$$H_1 T_d = 0_{(n\times 1)} \qquad [1.92]$$

$$H_{1_d} T_d = 1. \qquad [1.93]$$

The augmented estimated state vector reads:

$$\widehat{x} = H_1 \widehat{z} + H_2 y \qquad [1.94]$$

$$\widehat{d} = H_{1_d} \widehat{z} + H_{2_d} y \qquad [1.95]$$

and can be used to monitor the disturbance through $\widehat{d}$.

It can easily be shown that the state-feedback dynamics reads:

$$\text{spec}(A_a - B_a K_{c_a}) = \text{spec}\left(\begin{bmatrix} A - BK_c & B - BK_d \\ 0 & \lambda \end{bmatrix}\right) = \text{spec}(A - BK_c) \cup \lambda$$

and that this dynamics (and so the closed-loop dynamics) is independent of the gain K_d. The applied control signal is $u = e - K_c\widehat{x} - K_d\widehat{d}$. Thus, choosing $K_d = 1$ tends to cancel the input disturbance d (once the steady state of the estimator is reached) and such a tuning does not perturb the closed-loop dynamics. Although the closed-loop dynamics does not change, the controller is no longer the same. It can be shown that the tuning $K_d = 1$ leads to a new controller $K_d(s)$ with one eigenvalue assigned to λ, that is (from [1.65] considering that $D = 0$ and $D_Q = 0$):

$$\boxed{\lambda \in \text{spec}\,(F - T_a B_a K_{c_a} H_{1_a}).}$$

PROOF.– Let A_{K_d} be the new controller state matrix:

$$A_{K_d} = F - T_a B_a K_{c_a} H_{1_a} \tag{1.96}$$

$$= F - [T \;\; T_d] \begin{bmatrix} B \\ 0 \end{bmatrix} [K_c \;\; 1] \begin{bmatrix} H_1 \\ H_{1_d} \end{bmatrix} \tag{1.97}$$

$$= F - TB(K_c H_1 + H_{1_d}) \tag{1.98}$$

$$= F - (F - \lambda I_{n_K}) T_d (K_c H_1 + H_{1_d}) \quad \text{with [1.89]} \tag{1.99}$$

$$= (F - \lambda I_{n_K})(I_{n_K} - T_d(K_c H_1 + H_{1_d})) + \lambda I_{n_K}. \tag{1.100}$$

Then:

$$\begin{aligned} A_{K_d} T_d &= (F - \lambda I_{n_K})(T_d - T_d(K_c H_1 T_d + H_{1_d} T_d)) + \lambda T_d \\ &= (F - \lambda I_{n_K})(T_d - T_d(0 + 1)) + \lambda T_d \text{ with [1.92] and [1.93]} \\ &= \lambda T_d. \end{aligned}$$

Thus, T_d is a right eigenvector of A_{K_d} associated with the eigenvalue λ.

A quite interesting case is $\lambda = 0$, which can be used to model a constant input disturbance (input bias). Then, to reject this constant disturbance, the new controller exhibits an integrator. Such a behavior is well-known from a practical point of view. Thus, given an initial stabilizing controller $K_0(s)$ and a model $G_0(s)$, it is possible to design a new controller $K_d(s)$ with an integral term to reject a constant input disturbance. The closed-loop dynamics obtained with $K_d(s)$ is the same as the dynamics obtained with $K_0(s)$ and thus is stable. That is detailed in the following illustration.

1.10.2. *Illustration*

The model $G_0(s) = 1/(s^2 - 1)$ of the launcher and the controller $K_3(s) = -(50s + 70)/(s^2 + 10s + 50)$ are still considered for this application. The external disturbance $d(t)$ acts on the plant input according to Figure 1.23. For simulation analyzes, the signal $d(t)$ is a bias (0.2 rad) with a small time-domain drift (−0.002 rad/s), that is:

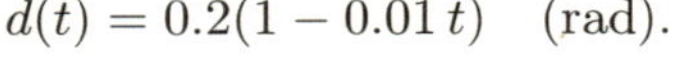

$$d(t) = 0.2(1 - 0.01\,t) \quad \text{(rad)}.$$

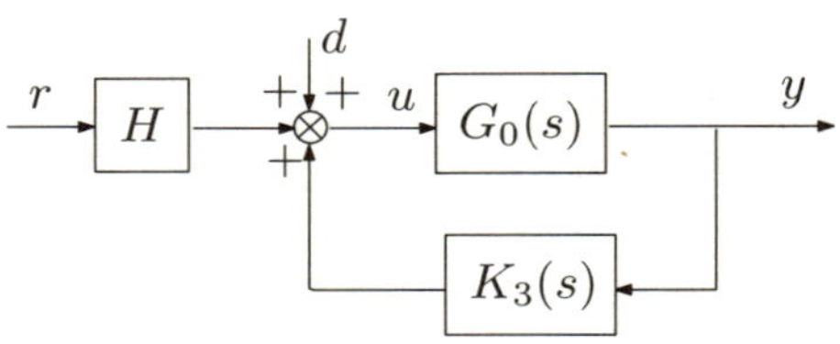

Figure 1.23. *Positive feedback connection of plant $G_0(s)$ and controller $K_3(s)$ with reference input $r(t)$, static-feedforward H and input disturbance $d(t)$*

Considering the servo-loop of Figure 1.23 (with $H = 0.4$, see section 1.9.2), the response of $y = \alpha(t)$ to a square signal on $r(t) = \alpha_{ref}(t)$ (i.e. a reference input on the angle of attack), to initial conditions on launcher states ($\alpha(t = 0) = 1$ rad and $\dot{\alpha}(t = 0) = -1$ rad/s) and to the input disturbance $d(t)$ is depicted in Figure 1.24. This response, in comparison with Figure 1.19 (without disturbance), exhibits a large steady-state error obviously due to the input disturbance $d(t)$.

The model of the disturbance is an unknown constant bias, that is $\lambda = 0$ in equation [1.87] and the augmented model $G_a(s) = D + C_a(sI - A_a)^{-1}B_a$ associated with the augmented state vector $x_a = [\alpha \;\; \dot{\alpha} \;\; d]^T$ reads:

$$\begin{bmatrix} \dot{\alpha} \\ \ddot{\alpha} \\ \dot{d} \\ \hline \alpha \end{bmatrix} = \left[\begin{array}{ccc|c} 0 & 1 & 0 & 0 \\ 1 & 0 & 1 & 1 \\ 0 & 0 & 0 & 0 \\ \hline 1 & 0 & 0 & 0 \end{array}\right] \begin{bmatrix} \alpha \\ \dot{\alpha} \\ d \\ \hline \delta \end{bmatrix}.$$

The observer-based realization (T_a, F and G) of the second-order controller $K_3(s)$ on the third-order augmented model $G_a(s)$ is then computed using the function `cor2tfg` (see section A4.1, Appendix 4) choosing to assign the fast dynamics (6.36 rad/s) to the state-feedback dynamics. The parameters K_{c_a}, H_{1_a} and H_{2_a} ($D_Q = 0$) are then computed using formulas [1.67] and [1.68]. The observer-based controller is then implemented according to Figure 1.25 where the augmented reference state is $x_{e_a} = [\alpha_{\text{ref}} \;\; 0 \;\; 0]^T$ and the corresponding equilibrium input is $u_e = -\alpha_{\text{ref}}$.

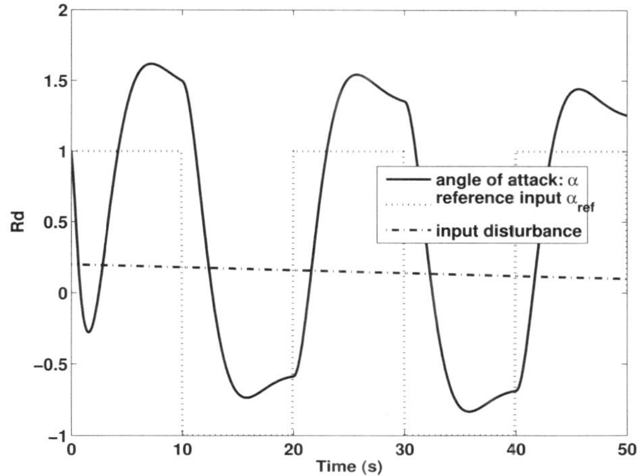

Figure 1.24. *Time-domain responses of classical loop (see Figure 1.23)*

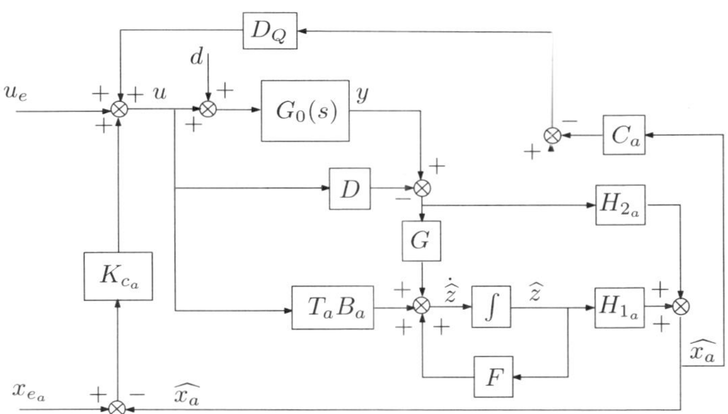

Figure 1.25. *Observer-based controller on the augmented model* $G_a(s) = D + C_a(sI - A_a)^{-1}B_a$ *with equilibrium state* x_{e_a} *and input* u_e *as reference inputs*

Figure 1.26 illustrates the time-domain responses of the reference input $\alpha_{\text{ref}}(t)$, the input disturbance $d(t)$, the plant output $y(t) = \alpha(t)$ and the disturbance estimation $\widehat{x_a(3)} = \widehat{d}$. Although the steady-error on $y(t)$ is still not satisfactory, we can note that the disturbance estimation is correct and converges to the actual input disturbance $d(t)$ according to the estimation dynamics (here 0.703 rad/s). Such an analysis can be performed using the following Matlab® sequence[5]:

5 The reader is advised to use a SIMULINK file based on Figure 1.25 to perform such analyzes.

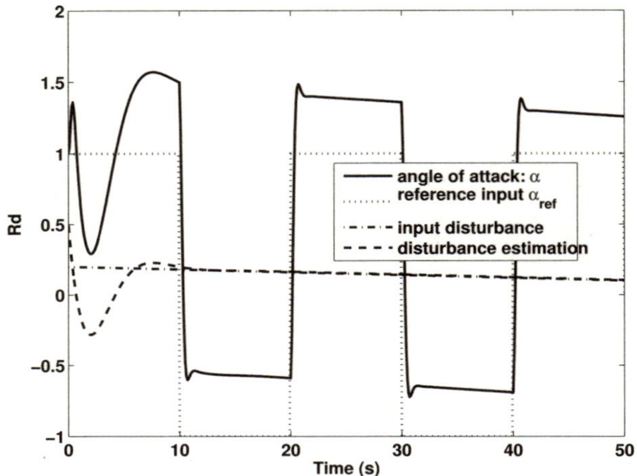

Figure 1.26. *Time-domain responses using observed-based realization (fast dynamics* (6.36 *rad/s) is assigned to spec*$(A_a - B_a K_{c_a})$*)*

```
>> % Illustration : External disturbance monitoring and rejection:
>> G0=ss([0 1;1 0],[0;1],[1 0],0); % Nominal plant
>> G0d=G0*[1 1];                   % Plant with input disturbance
>> % Time-domain response of external signals:
>> Time=[0:0.1:50];
>> U=zeros(size(Time));
>> U([1:100 201:300 401:500])=1;
>> U([101:200 301:400])=-1;        % Reference input signal
>> Dis=[0.2:-0.1/500:0.1];         % Actual disturbance
>> % Initial controller:
>> K3=-tf([50 70],[1 10 50]);
>> [A,B,C,D]=ssdata(G0);
>> % Augmented model with an unknown bias on the input:
>> Ga=ss([A B;zeros(1,3)],[B;0],[C 0],D);
>> K3ss=ss(K3);
>> % Computing the observer-based realization
>> [Ta,F,G]=cor2tfg(Ga,K3ss);      % Affect the slow dynamics to F !!
>> damp(F)                         % Estimation dynamics

        Eigenvalue                  Damping       Freq. (rad/s)

  -4.38e-01 + 5.49e-01i           6.24e-01          7.03e-01
  -4.38e-01 - 5.49e-01i           6.24e-01          7.03e-01

>> H1H2=inv([Ta;Ga.c]);
>> H1a=H1H2(:,1:2);H2a=H1H2(:,3);
>> Jm=inv(eye(size(Ga.b,2))-K3ss.d*Ga.d);
>> Kca=-Jm*(K3ss.c*Ta+K3ss.d*Ga.c);
>> DQ=0;
>> damp(Ga.a-Ga.b*Kca)             % Augmented state-feedback dynamics
```

```
          Eigenvalue                  Damping        Freq. (rad/s)

 -4.56e+00 + 4.44e+00i            7.17e-01          6.36e+00
 -4.56e+00 - 4.44e+00i            7.17e-01          6.36e+00
  0.00e+00                       -1.00e+00          0.00e+00

>> % 2 d.o.f: observer-based controller with the disturbance estimation
>> % on the second output:
>> Kbar=ss(F-Ta*Ga.b*(Kca+DQ*Ga.c)*H1a,...
          [G+Ta*Ga.b*(DQ-(Kca+DQ*Ga.c)*H2a) Ta*Ga.b],...
          [-(Kca+DQ*Ga.c)*H1a;H1a(3,:)],...
          [DQ-(Kca+DQ*Ga.c)*H2a 1;H2a(3,:) 0]);
>> Kobs=feedback(Kbar,Ga.d,1,1);
>> % Taking into account the reference input: e=u_e+K_c*x_e:
>> Kobs=Kobs*[1 0;0 Kca(1)-1];
>> % Closed-loop system:
>> BF=connect(append(G0d,Kobs),[2 2;3 1],[1 4],[1 3]);
>> [y,t,x]=lsim(BF,[Dis' U'],Time,[1 -1 0 0]);
>> figure
>> plot(t,y(:,1),'k-','LineWidth',2)
>> hold on
>> plot(t,U,'k:','LineWidth',2)
>> plot(t,Dis,'k-.','LineWidth',2)
>> plot(t,y(:,2),'k--','LineWidth',2)
>> xlabel('Time (s)');
>> ylabel('Rd');
>> [LEGH1,OBJH1]=legend('angle of attack: \alpha',...
              'reference input \alpha_{ref}','input disturbance',...
              'disturbance estimation');
```

To reject the disturbance, the gain on $\widehat{d}$ is set to 1. That is, $K_{c_a}(3) = 1$. The new time-domain responses are then depicted in Figure 1.27. It can be noted that after a transient response due to the estimation dynamics, the input reference tracking and the disturbance rejection are correct. The integral term in the new controller and the closed-loop dynamics, exactly assigned to the initial one, are highlighted by the following complementary Matlab® sequence:

```
>> % Disturbance rejection:
>> Kca(3)=1;
>> Kbar=ss(F-Ta*Ga.b*(Kca+DQ*Ga.c)*H1a,...
          [G+Ta*Ga.b*(DQ-(Kca+DQ*Ga.c)*H2a) Ta*Ga.b],...
          [-(Kca+DQ*Ga.c)*H1a;H1a(3,:)],...
          [DQ-(Kca+DQ*Ga.c)*H2a 1;H2a(3,:) 0]);
>> Kobs=feedback(Kbar,Ga.d,1,1);
>> % Taking into account reference input: e=u_e+K_c*x_e:
>> Kobs=Kobs*[1 0;0 Kca(1)-1];
>> % Integral effect:
>> zpk(Kobs(1,1))
```

```
Zero/pole/gain:
-50 (s^2 + s + 0.4)
--------------------
     s (s+10)

>> % Closed-loop system:
>> BF=connect(append(G0d,Kobs),[2 2;3 1],[1 4],[1 3]);
>> % But closed loop eigenvalues are the same:
>> damp(BF)

        Eigenvalue              Damping       Freq. (rad/s)

  -4.38e-01 + 5.49e-01i        6.24e-01         7.03e-01
  -4.38e-01 - 5.49e-01i        6.24e-01         7.03e-01
  -4.56e+00 + 4.44e+00i        7.17e-01         6.36e+00
  -4.56e+00 - 4.44e+00i        7.17e-01         6.36e+00
```

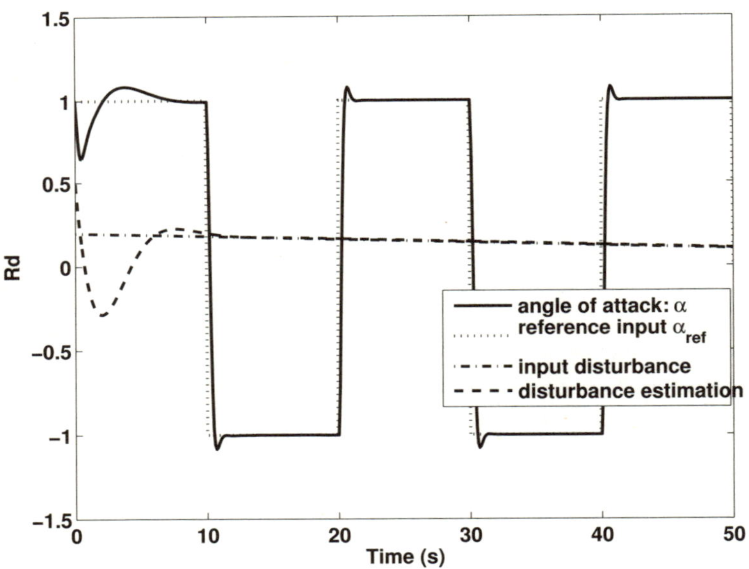

Figure 1.27. *Time-domain responses using observed-based realization (fast dynamics (*6.36 *rad/s) is assigned to spec*$(A_a - B_a K_{c_a})$*) with* $K_c(3) = 1$ *(integral effect)*

1.11. Minimal parametric description of a linear system

Observer-based realization of a given system can be used to find a representation with a minimal number of parameters (descriptive coefficients). Indeed, in the general case of a nth-order system $G(s)$ with m inputs and p outputs (with $n \geq m$ and

$n \geq p$), the minimal number of parameters required to describe this system is $n(m+p)+pm$ while an arbitrary state-space representation involves $n^2+n(m+p)+pm$ parameters (i.e. the number of coefficients in matrices A, B, C and D). $G(s)$ can be seen as an observer-based controller on a p inputs – m outputs fictitious canonical plant $G_{\text{canon}}(s)$: $G_{\text{canon}}(s)$ is chosen as a set of n integrators in series connection where:

– the inputs of the first p integrators are connected to the p inputs;

– the outputs of the last m integrators are connected to the m outputs.

$G_{\text{canon}}(s)$ is described in Figure 1.28 and its state-space representation (equation [1.101]) involves only 0 and 1.

$$\dot{x} = \begin{bmatrix} 0_{(n-1)\times 1} & I_{n-1} \\ 0 & 0_{1\times(n-1)} \end{bmatrix} x + \begin{bmatrix} 0_{(n-p)\times p} \\ I_p \end{bmatrix} u = A_c x + B_c u \qquad [1.101]$$

$$y = \begin{bmatrix} I_m & 0_{m\times(n-m)} \end{bmatrix} x + 0_{m\times p} u = C_c x + D_c u.$$

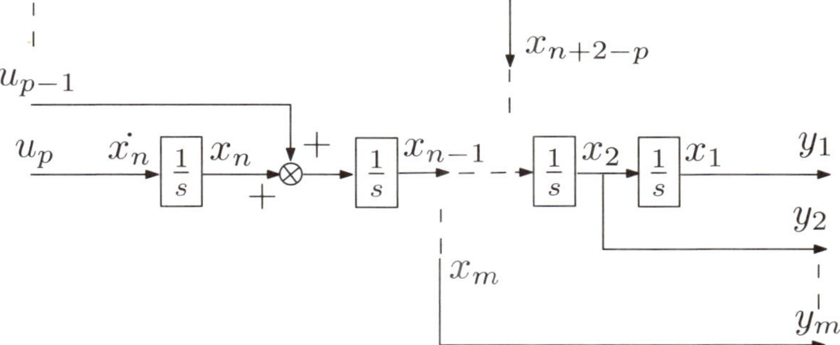

Figure 1.28. *Canonical model $G_{canon}(s)$ for a n-order, m inputs and p-outputs system $G(s)$ ($n \geq m$ and $n \geq p$)*

Then, $G(s)$ is entirely defined by the state-feedback gain K_c ($p \times n$ coefficients), the state-estimator gain K_f ($n \times m$ coefficients) and the static Youla parameter D_Q ($p \times m$ coefficients) according to the structure depicted in Figures 1.29 and 1.30. If $G(s)$ represents a reduced nth-order controller to be designed using fixed-order synthesis solvers now available [GAH 11, GUM 09], such a minimal parametric description can be useful to reduce the number of decision variables in the optimization process. The structure depicted in Figure 1.30 can then be used to simplify the problem as the design of a static gain, structured according to K_c, K_f and D_q.

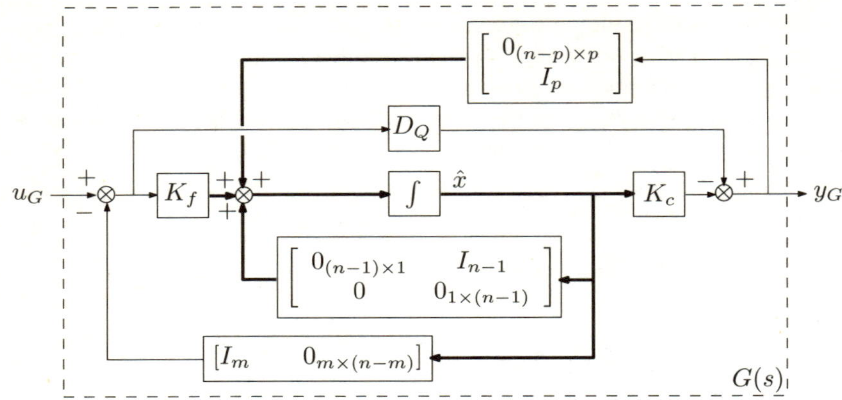

Figure 1.29. *Canonical observer-based representation of $G(s)$ between its m inputs u_G and its p outputs y_G*

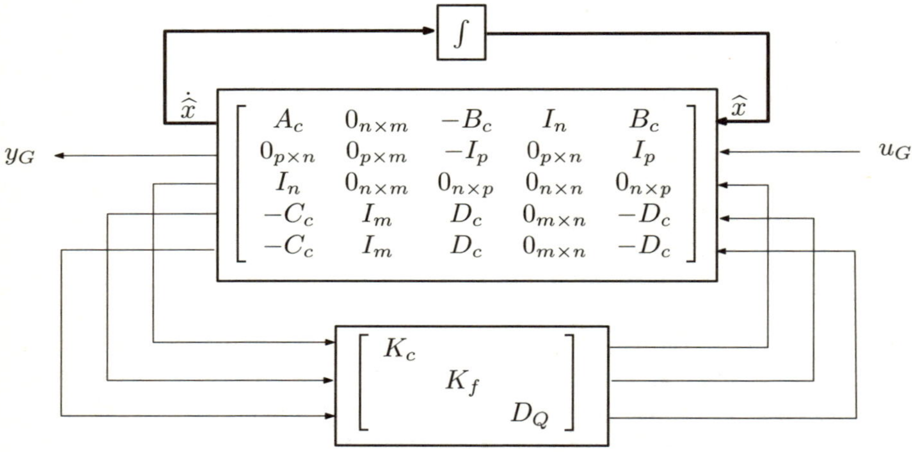

Figure 1.30. *Observer-based LFT representation of $G(s)$ between its m inputs u_G and its p outputs y_G as an observer-based structure on the model (A_c, B_c, C_c, and D_c) (defined by equation* [1.101], *for instance)*

The toolbox `bib_obr`[6] includes a Matlab® function `obcanon` to compute such a minimal parametric representation for a given system $G(s)$ (the `help` of this function can be found in section A4.5, Appendix 4).

6 See http://personnel.isae.fr/daniel-alazard/matlab-packages.

1.12. Selection of the observer-based realization

Section 1.5 highlighted the combinatory set of solutions in the computation of the observer-based realization of a given controller and a given plant. The solution only depends:

– on the way the $n+n_K$ closed-loop eigenvalues are split between the n_K eigenvalues of the Luenberger observer dynamics (spec(F)) and the n remaining eigenvalues assigned to the control dynamics;

– and second (if $n_K > n$), on the way the n_K eigenvalues of F are split between the n eigenvalues of the state-estimator dynamics (spec$(A - K_fC)$) and the $n_K - n$ eigenvalues of the Youla parameter dynamics (spec(A_Q)).

The Matlab® functions `cor2tfg` and `cor2obr` are interactive functions to help the user to manage these eigenvalue assignments using a graphic interface. Remark 1.2 gives some recommendations in the assignment strategies based on practical considerations which were embedded in Matlab® functions detailed in the following two sections.

1.12.1. *Luenberger observer dynamics assignment*

In order to respect the dynamic behavior of the physical plant and reduce the state-feedback gains, it is recommended to assign to the control dynamics the n closed-loop eigenvalues which are the "nearest" from the n plant eigenvalues. The remaining n_K closed-loop eigenvalues are then assigned to the Luenberger observer dynamics (spec(F)). Such a systematic assignment can easily be done tracking the n plant eigenvalues from the open-loop to the closed-loop and using a root locus-like procedure. With this objective in mind, such a procedure must increment a single loop gain k acting on all the m control channels from 0 to 1 with small steps (see Figure 1.31) and isolate at each step the n eigenvalues starting from the n plant ($G_0(s)$) eigenvalues in a least-squares sense. The Matlab® function `cor2tfga` in the toolbox `bib_obr` computes the matrices T, F and G from a given controller $K_0(s)$ and a given plant $G_0(s)$ using such an iterative search and taking care:

– to manage bifurcations which may occur on such a root locus in order to find a real parametrization if possible (see remark 1.2);

– to assign together auto-conjugate pairs of eigenvalues;

– to ensure that plant uncontrollable eigenvalues are not assigned to Luenberger observer dynamics.

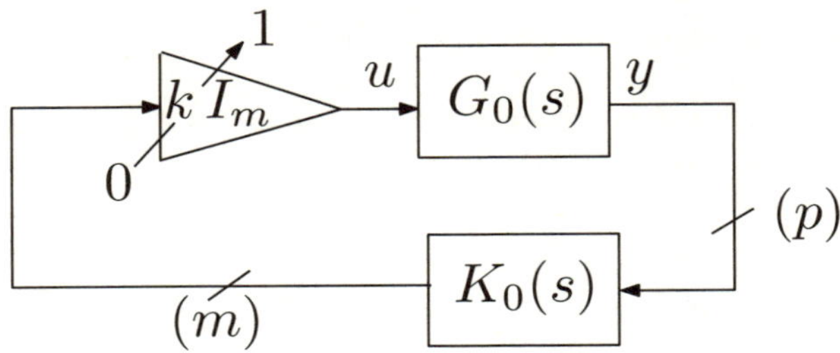

Figure 1.31. *Closed-loop with a loop gain on the m control signals varying from 0 (open-loop) to 1 (nominal closed-loop)*

Contrary to the function `cor2tfg`, this function `cor2tfga` is completely autonomous and does not require the intervention of the user. The `help` of this function can be found in section A4.6 (Appendix 4).

1.12.2. *State-estimator dynamics assignment*

In the case $n_K > n$, it can also be recommended to assign the $n_K - n$ fastest eigenvalues of F to the Youla parameter dynamics spec(A_Q) in such a way that the Youla parameter $Q(s)$ acts as a direct feedthrough in the controller and can be eventually reduced *a posteriori*. In the same spirit of the function `cor2tfga`, the function `cor2obra` was created to implement such a rule and to provide (by itself) an observer-based parametrization that is: the state-feedback gain K_c, the state-estimator gain K_f and the Youla parameter $Q(s)$ of a given controller $K_0(s)$ for a given plant $G_0(s)$.

Note that the solution proposed by the function `cor2obra` can be analyzed using the function `obrmap` from the dynamics assignment point of view: this function plots in the (s or z) complex plane the state-feedback dynamics, the state-estimation dynamics and the Youla parameter dynamics for a given observer-based parametrization relative to a given plant. Open-loop plant and controller dynamics are also plotted to appreciate the closed-loop behavior.

The `help` of these functions, included in the toolbox `bib_obr`, can be found in sections A4.7 and A4.8 (Appendix 4).

The file named `demo_obr.m` also illustrates how to use these functions to compute an observer-based realization of the 16th-order controller designed in the demo #3 of the `Mu-Analysis and Synthesis Toolbox`.

1.13. Conclusions

In this chapter, a procedure to compute a minimal observer-based realization of an arbitrary controller was proposed. This technique is based on the resolution of a generalized non-symmetric Riccati equation. Necessary conditions were given for the solvability of this equation in terms of observability and controllability properties of the plant. The interest of observer-based realization for gain scheduling, controller switching, state and disturbance monitoring and reference input tracking was highlighted on a very simple example. Demo files are available for readers who wish to practice.

Further work is still needed to exploit the multiplicity of choices in the distribution of the closed-loop eigenvalues between the state-feedback eigenvalues, the state-estimator eigenvalues and the Youla parameter eigenvalues. This problem is particularly important to smoothly interpolate or schedule a family of state-feedback gains and state-estimator gains for practical problems requiring some gain-scheduling strategy. The usefulness of these controller structures to handle input saturation constraints also deserves further development.

The reader will find in Voinot *et al.* [VOI 03] an application of this technique to implement gain-scheduled controllers for the attitude control of a launcher. In Alazard and Apkarian [ALA 07], the observer-based realization of an aircraft flight control law is used to monitor on-line the disturbance (gust). For this application, the on-board model used to compute the observer-based realization includes a model of the disturbance.

The capability of observer-based structures to monitor plant states or disturbances opens a new possibility to update a given initial controller according to the actual plant behavior and allows some links with adaptive approaches to be imagined. This capability to mix various approaches is the main objective of the reverse engineering presented in this book.

1.14. Bibliography

[ALA 99] ALAZARD D., APKARIAN P., "Exact observer-based structures for arbitrary compensators", *International Journal of Robust and Non-Linear Control*, Montreal (CA), vol. 9, pp. 101–118, 1999.

[ALA 01] ALAZARD D., "Extracting physical tuning potentiometers from a complex control law: application to flexible aircraft flight control", *AIAA, Guidance, Navigation and Control Conference*, Montreal, Canada, August 2001.

[ALA 07] ALAZARD D., APKARIAN P., "Disturbances monitoring from controller states", *Proceedings of the SSSC'07, IFAC Symposium on Systems Structure and Control*, Foz do Iguasu, Brazil, October 2007.

[BEN 85] BENDER D.J., FOWELL R.A., "Computing the estimator-controller form of a compensator", *International Journal of Control*, vol. 41, pp. 1565–1575, 1985.

[BEN 86] BENDER D.J., FOWELL R.A., ASSAL A. F., "Estimating the plant state from the compensator state", *IEEE Transactions on Automatic Control*, vol. 31, pp. 964–967, 1986.

[BOY 91] BOYD S., BARRATT G., *Linear Controller Design: Limit of Performance*, Prentice-Hall, 1991.

[CUM 04] CUMER C., DELMOND F., ALAZARD D., CHIAPPA C., "Tuning of observer-based controllers", *Journal of Guidance, Control and Dynamics*, vol. 27, no. 4, pp. 607–615, 2004.

[FRA 81] FRANKLIN G.F., JR., JOHNSON C.R. J., "A condition for full zero assignment in linear control systems", *IEEE Transactions on Automatic Control*, vol. AC-26, no. 2, pp. 519–521, 1981.

[GAH 11] GAHINET P., APKARIAN P., "Structured H_∞ synthesis in MATLAB", *Proceedings of the 18th IFAC World Congress*, Milan, Italy, pp. 1435–1440, 28 August–2 September 2011.

[GOL 96] GOLUB G.H., VAN LOAN C.F., *Matrix - Computations*, *Studies in the Mathematical Sciences*, J. Hopkins University Press, Baltimore, MD. 1996.

[GUM 09] GUMUSSOY S., HENRION D., MILLSTONE M., OVERTON M., "Multiobjective Robust Control with HIFOO 2.0", *Proceedings of the IFAC Symposium on Robust Control Design,* Haïfa, Israel, June 2009.

[KAM 95] KAMINER I., PASCOAL A.M., KHARGONEKAR P.P., COLEMAN E.E., "A velocity algorithm for the implementation of gain-scheduled controllers", *Automatica*, vol. 31, no. 8, pp. 1185–1191, 1995.

[LAW 95] LAWRENCE D.A., RUGH W.J., "Gain scheduling dynamic linear controllers for a nonlinear plant", *Automatica*, vol. 31, no. 3, pp. 381–290, 1995.

[LUE 71] LUENBERGER D.G., "An introduction to observers", *IEEE Transactions on Automatic Control*, vol. AC-16, pp. 596–602, 1971.

[NAR 77] NARASIMHA-MURTHI N., WU F.F., "On the Riccati equation arising from the study of singularly perturbed systems", *IEEE Joint Automatic Control Conference*, New York, 22–24 June 1977.

[PEL 00] PELLANDA P.C., APKARIAN P., ALAZARD D., "Gain-scheduling through continuation of observer-based realizations – applications to $\mathcal{H}_\infty$ and μ controllers", *Proceedings of the 39th IEEE Conference on Decision and Control*, Sydney, IEEE, pp. 2787–2792, 12–15 December 2000.

[STI 99] STILWELL D.J., RUGH W., "Interpolation of observer state feedback controllers for gain scheduling", *IEEE Transactions on Automatic Control*, vol. 44, no. 6, pp. 1225–1229, 1999.

[TAR 97] TARBOURIECH S., GARCIA G., *Control of uncertain systems with bounded inputs*, Lectures Notes in Control and Information Sciences, Springer Verlag, 1997.

[VOI 03] VOINOT O., ALAZARD D., APKARIAN P., MAUFFREY S., CLÉMENT B., "Launcher attitude control: discrete-time robust design and gain-scheduling", *Control Engineering Practice*, vol. 11, pp. 1243–1252, 2003.

[ZHO 96] ZHOU K., DOYLE J. C., GLOVER K., *Robust and Optimal Control*, Prentice Hall, 1996.

2

Cross Standard Form and Reverse Engineering

2.1. Introduction

In most practical applications, the control design problem can be expressed in the following terms: is it possible to improve a given controller (often, a simple low-order controller designed upon a particular know-how or good sense rules) to meet additional H_2 or H_∞ specifications? or in other terms: is it possible to take into account a given controller (which meets some closed-loop specifications) in a standard H_2 and H_∞ control problem? To address this *reverse engineering* problem, the notion of the Cross Standard Form (CSF) is introduced in this section for a given nth-order plant and an arbitrary given stabilizing n_Kth-order controller [DEL 06]. The CSF can be seen as a solution for both inverse H_∞- and H_2-optimal control problems, that is: the CSF is a standard augmented problem whose *unique* H_∞- or H_2-optimal controller is an arbitrary given controller. The CSF is directly defined by the four state-space matrices of the plant, the four state-space matrices of the given controller and the solution T of the general non-symmetric Riccati equation [1.18] introduced in Chapter 1 to compute the observer-based realization of a given controller for a given plant. The CSF can be applied to full-order, low-order or augmented-order controllers.

The interest for inverse optimal control problems motivates many works [KAL 64, MOL 73, FUJ 87, LEN 88, FUJ 88, SEB 01]. The practical interest of such solutions lies in the possibility to mix various approaches or take into account different kinds of specifications [SUG 87, SUG 98, SHI 97]. In the particular case of the H_∞-optimal control problem, the various contributions address restrictive cases: the state-feedback controller in [FUJ 88], a single-output controller and specific sensitivity problem in [LEN 88]. But a solution for the general case (multi-input multi-output, dynamic output feedback of arbitrary order) has never been stated. This general case is addressed in [SEB 01]: for a given weight system $W(s)$, a given controller $K(s)$ and a given positive scalar γ, the problem is to find all the plants $G(s)$ such that $\|F_l(F_l(W,G),K)\|_\infty < \gamma$ (see Figure 2.1). Note that the problem considered in this section is different since the plant $G(s)$ (i.e. the lower right-hand transfer matrix of the standard augmented plant $P = F_l(W,G)$) is given and corresponds to the model of the plant between the control input u and the measured output y.

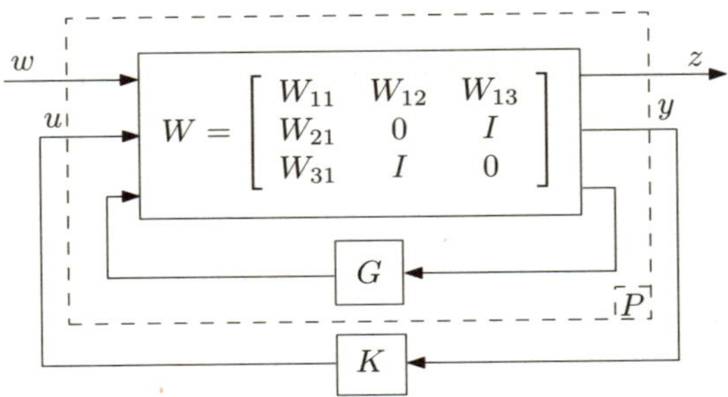

Figure 2.1. *Block diagram of standard plant P, weight function W, model G and controller K*

Convex optimization [BOY 91] seems also an attractive approach to take into account a given controller and additional H_2 or H_∞ constraints. But such an approach needs a Youla parametrization of the controller and so is limited to full-order (observer-based) controllers. Furthermore, this approach leads to very high order controllers. More recently, CSF is used to initialize a Model Predictive Control (MPC) design procedure in [HAR 11]. Authors of this reference are at the origin of the term “reverse engineering” that has been adopted in this book.

In section 2.2, the CSF is defined as a solution to H_2 and H_∞ inverse optimal control problems, for an nth-order Linear Time Invariant (LTI) system and an n_Kth-order stabilizing LTI controller. In section 2.3, an analytical expression of a CSF is proposed for low-order controllers ($n_K \leq n$) and the existence of such a CSF is discussed. It will be shown that the unique optimal controller provided by full-order H_∞ controller (convex) solvers is non-minimal and is equivalent to the given low-order controller from the input-output behavior. Today, new H_∞ synthesis solvers allow fixed-structure, and so reduced-order, controllers to be directly designed [APK 06, GAH 11, ARZ 11, GUM 09, GAB 10, GAB 12]. Thus, the inverse optimal H_∞ control problem for reduced-order controller is now solved. In section 2.4, the CSF is extended for augmented-order controllers ($n_K > n$) and so encompasses previous results presented in [ALA 04]. Finally, the second-order model of a launcher is used in section 2.5 to highlight the way to use CSF in order to take into account an initial low-order controller and a frequency-domain specification. It will be also shown how a basic initial controller fitting a prescribed dominant dynamics can be augmented with a phase lead term to cope with a too slow actuator dynamics.

2.2. Definitions

The general standard plant between exogenous input w, control input u, controlled output z and measurement output y is denoted as:

$$P(s) = \begin{bmatrix} P_{zw}(s) & P_{zu}(s) \\ P_{yw}(s) & P_{yu}(s) \end{bmatrix},$$

with the corresponding state space realization:

$$P(s) := \left[\begin{array}{c|cc} A_p & B_1 & B_2 \\ \hline C_1 & D_{11} & D_{12} \\ C_2 & D_{21} & D_{22} \end{array}\right]. \qquad [2.1]$$

Let us consider again the plant $G_0(s)$ defined in [1.2] and the stabilizing initial controller $K_0(s)$ defined in [1.7].

DEFINITION 2.1.– Inverse H_2-optimal problem
Find a standard plant $P(s)$ such that:

– $P_{yu}(s) = G_0(s)$;

– K_0 *stabilizes* $P(s)$;

– $K_0(s) = \arg\min_{K(s)} \|F_l(P(s), K(s))\|_2$;

(namely: $K_0(s)$ *minimizes* $\|F_l(P(s), K(s))\|_2$*).*

DEFINITION 2.2.– Inverse H_∞-optimal problem
Find a standard plant $P(s)$ such that:

– $P_{yu}(s) = G_0(s)$;

– K_0 *stabilizes* $P(s)$;

– $K_0(s) = \arg\min_{K(s)} \|F_l(P(s), K(s))\|_\infty$.

DEFINITION 2.3.– Cross Standard Form
If the standard plant $P(s)$ is such that the four conditions:

– *C1:* $P_{yu}(s) = G_0(s)$;

– *C2:* K_0 *stabilizes* $P(s)$;

– *C3:* $F_l(P(s), K_0(s)) = 0$;

– *C4:* K_0 *is the unique solution of the optimal* H_2 *or* H_∞ *problem* $P(s)$,

are met, then $P(s)$ *is called the Cross Standard Form (CSF) associated with the system* $G_0(s)$ *and the controller* $K_0(s)$ *and will be denoted* $P_{CSF}(s)$ *in the sequel.*

By construction, the CSF solves the inverse H_2-optimal problem and the inverse H_∞-optimal problem. Note that the uniqueness condition C4 is relevant in our context since we are looking for an H_2 or H_∞ design to recopy a given controller.

2.3. Low-order controller case ($n_K \leq n$)

The following proposition provides a general analytical characterization of the CSF.

PROPOSITION 2.1.– *For a given stabilizable and detectable nth-order system* $G_0(s)$ *(equation* [1.2]*) and a given stabilizing* n_K*th-order controller* $K_0(s)$ *with* $n_K < n$ *(equation* [1.7]*), a CSF reads:*

$$P_{CSF}(s) := \left[\begin{array}{c|cc} A & T^{\#}B_K - BD_K & B \\ \hline -C_KT - D_KC & D_KDD_K - D_K & I_m - D_KD \\ C & I_p - DD_K & D \end{array}\right], \qquad [2.2]$$

where T *is the solution of the generalized Riccati equation* [1.18] *and where* $T^{\#}$ *is a right inverse*[1] *of* T *(such that* $TT^{\#} = I_{n_K}$*).*

PROOF.– From [2.2], it is obvious that conditions C1 and C2 are met. A state-space realization of $F_l(P_{CSF}, K_0)$ associated with state vector $[x^{\mathrm{T}}, \quad x_K^{\mathrm{T}}]^{\mathrm{T}}$ reads:

$$\left[\begin{array}{cc|c} A + BJ_mD_KC & BJ_mC_K & T^{\#}B_K \\ B_KJ_pC & A_K + B_KDJ_mC_K & B_K \\ \hline -C_KT & C_K & 0 \end{array}\right],$$

where J_m and J_p are defined in [1.8]. Let us consider the change of state coordinates (already defined in equation [1.16]):

$$\mathcal{M} = \mathcal{M}^{-1} = \begin{bmatrix} I_n & 0 \\ T & -I_{n_K} \end{bmatrix}, \qquad [2.3]$$

1 See also proposition 2.2.

where T is a solution of [1.18] and $TT^{\#} = I_{n_K}$. The new state space realization of $F_l(P_{CSF}, K_0)$ reads:

$$\left[\begin{array}{cc|c} A + BJ_m(D_K C + C_K T) & -BJ_m C_K & T^{\#} B_K \\ 0 & A_K + (B_K D - TB)J_m C_K & 0 \\ \hline 0 & -C_K & 0 \end{array}\right]. \qquad [2.4]$$

So the $n + n_K$ (stable) close-loop eigenvalues are composed of:

– n eigenvalues of $A + BJ_m(D_K C + C_K T)$, which are unobservable by the controlled output z of P_{CSF},

– n_K eigenvalues of $A_K + (B_K D - TB)J_m C_K$, which are uncontrollable by the exogenous input w of P_{CSF}.

Thus, condition C3 is met:

$$F_l(P_{CSF}(s), K_0(s)) = 0.$$

In the next section, it is shown that it is always possible to find a right-inverse $T^{\#}$ of T such that the uniqueness condition C4 is met and that ends the proof.

The general block-diagram representation of P_{CSF} is depicted in Figure 2.2. We can note that the CSF is a one-block problem and can be seen as a combination of a Output Estimation (OE) problem and a Disturbance Feed-forward (DF) problem [ZHO 96]. So, if both cross transfers ($P_{zu}(s)$ and $P_{yw}(s)$) are minimum phase (no zero in the closed right half plane), then both H_2 and H_∞ syntheses converge toward the same H_∞ performance index (γ) [ZHO 92]. But for the standard problem P_{CSF}, we can state that $\gamma = 0$ and that both syntheses are exactly equal.

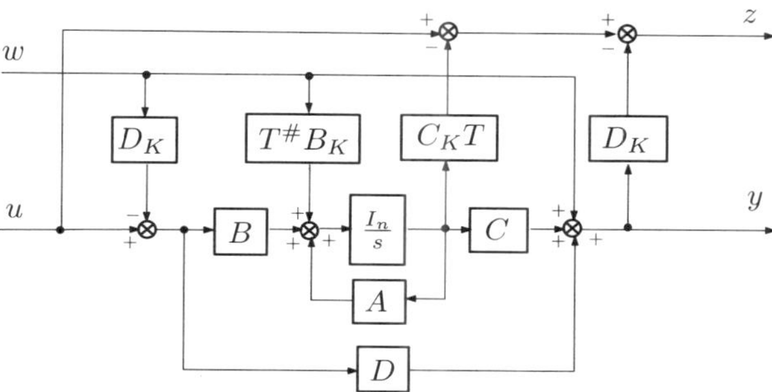

Figure 2.2. *Block diagram of Cross Standard Form $P_{CSF}(s)$ (case $n_K \leq n$)*

2.3.1. *Uniqueness condition*

The uniqueness condition (C4) can be proved considering the H_2-optimal controller of P_{CSF}: first, to vanish the direct feed-trough between exogenous inputs and controlled outputs in P_{CSF}, a simple change of variable ($u \leftarrow u - D_K y$) is performed to transform P_{CSF} into the problem $\overline{P_{CSF}}(s)$:

$$\left[\begin{array}{c|cc} A + BJ_m D_K C & T^{\#} B_K & BJ_m \\ \hline -C_K T & 0 & I_m \\ J_p C & I_p & DJ_m \end{array}\right], \qquad [2.5]$$

and thus:

$$F_l(P_{CSF}, K) = F_l(\overline{P_{CSF}}, K - D_K),$$
$$\arg\min_K \|F_l(P_{CSF}, K)\| = \arg\min_K \|F_l(\overline{P_{CSF}}, K)\| + D_K.$$

In [DOY 89] and [ZHO 96], it is demonstrated that a standard problem P has a unique H_2-optimal controller if and only if P is a regular problem. That is, in our case, if cross transfers:

$$P_{zu}(s){:} = \left[\begin{array}{c|c} A + BJ_m D_K C & BJ_m \\ \hline -C_K T & I_m \end{array}\right] \text{ and } P_{yw}(s){:} = \left[\begin{array}{c|c} A + BJ_m D_K C & T^{\#} B_K \\ \hline J_p C & I_p \end{array}\right]$$

have no invariant zeros on the $j\omega$ axis. It is clear that the n zeros of $P_{zu}(s)$ are the n eigenvalues of $\phi_{zu} = A + BJ_m(D_K C + C_K T)$ (ϕ_{zu} is the dynamic matrix of $P_{zu}^{-1}(s)$) and, considering [2.4], belong to the set of $n + n_K$ closed-loop eigenvalues and thus are stable by assumption. *So, $P_{zu}(s)$ has no zeros on the $j\omega$ axis.*

The problem of the zeros of $P_{yw}(s)$ is more complex: the n zeros of $P_{yw}(s)$ are the n eigenvalues of $\phi_{yw} = A + BJ_m D_K C - T^{\#} B_K J_p C$ (ϕ_{yw} is the dynamic matrix of $P_{yw}^{-1}(s)$). Then, pre-multiplying ϕ_{yw} by $N = [T^{\#} \quad T^{\perp}]$, post-multiplying by $N^{-1} = [T^{\mathrm{T}} \quad T^{\perp}]^{\mathrm{T}}$ and using [1.18], it comes:

$$N^{-1}\phi_{yw}N = \begin{bmatrix} A_K + (B_K D - TB)J_m C_K & 0 \\ \star & {T^{\perp}}^{\mathrm{T}}(A + BJ_m D_K C - T^{\#} B_K J_p C)T^{\perp} \end{bmatrix}.$$

The n zeros of $P_{yw}(s)$ are therefore composed of:

– n_K eigenvalues of $A_K + (B_K D - TB)J_m C_K$. Considering [2.4], these eigenvalues belong to the set of $n + n_K$ closed-loop eigenvalues and thus are stable by assumption;

– $n - n_K$ eigenvalues of $\varphi(T^{\#}) = {T^{\perp}}^{\mathrm{T}}(A + BJ_m D_K C - T^{\#} B_K J_p C)T^{\perp}$ whose location in the complex plane is discussed in the following proposition.

PROPOSITION 2.2.– *It is always possible to find a right-inverse $T^{\#}$ of T such that all the $n - n_K$ eigenvalues of $\varphi(T^{\#})$ (and thus all the n zeros of the cross transfer P_{yw}) are not on the $j\omega$ axis.*

PROOF.– The set of right-inverse matrices of T can be parametrized in the following way:

$$T^{\#} = T^{+} + T^{\perp}X,$$

where X is an $(n - n_K) \times n_K$ matrix of free parameters. Then:

$$\varphi(T^{\#}) = \varphi(X) = {T^{\perp}}^{\mathrm{T}}(A + BJ_m D_K C)T^{\perp} - XB_K J_p CT^{\perp}. \qquad [2.6]$$

So, X allows the $n - n_K$ eigenvalues of φ to be assigned in the s-plane. The computation of X is in fact an eigenvalue assignment problem by a state-feedback X^{T} on the pair $({T^{\perp}}^{\mathrm{T}}(A + BJ_m D_K C)^{\mathrm{T}}T^{\perp}, (B_K J_p CT^{\perp})^{\mathrm{T}})$.

So, the proposition 2.2 allows us to state that $P_{zu}(s)$ *has no zeros on the* $j\omega$ *axis.* Thus, $\overline{P_{CSF}}(s)$ is regular and $K_0(s)$ is the unique solution of the H_2-optimal problem P_{CSF}.

As $F_l(P_{CSF}, K_0) = 0$, all controllers solution of the H_∞-optimal problem are also solutions of the H_2-optimal problem. Thus, $K_0(s)$ is also the unique solution of the H_∞-optimal problem P_{CSF}.

2.3.2. *Existence of a CSF*

PROPOSITION 2.3.– *The non-existence of a full-row rank matrix T solution of the generalized non-symmetric Riccati equation* [1.18] *implies the non-existence of a CSF for $G_0(s)$ and $K_0(s)$.*

PROOF.– Let us assume that a regular CSF exists for the strictly proper stabilizing controller $K_0(s) - D_K$ and for the stabilizable and detectable modified system $\overline{G_0}(s)$ (such a change of variable is not restrictive):

$$\overline{G_0}(s) := \left[\begin{array}{c|c} A + BJ_m D_K C & BJ_m \\ \hline J_p C & DJ_m \end{array}\right].$$

Then, it is shown in [DOY 89] that the unique solution $\widehat{K_{H_2}}$ of the corresponding H_2-optimal problem involves a state-feedback gain K_c and a state-estimator gain K_f (according to the structure depicted in Figure 1.1 with $Q(s) = 0$). The nth-order state-space realization of such a controller associated with the state vector $\widehat{x}$ reads:

$$\widehat{K_{H_2}} : = \left[\begin{array}{c|c} A + BJ_m D_K C - BJ_m K_c - K_f J_p C + K_f DJ_m K_c & K_f \\ \hline -K_c & 0 \end{array}\right]. \quad [2.7]$$

As the solution is unique: $\widehat{K_{H_2}}(s) = K_0(s) - D_K$. Thus, the state-space realization [2.7] is non-minimal if $n_K < n$. Thus, a projection matrix $S_{n_K \times n}$ (full-row rank) exits such that: $x_K = S\widehat{x}$ and

$$\begin{aligned} S(A + BJ_m D_K C - BJ_m K_c - K_f J_p C + K_f DJ_m K_c) &= A_K S \\ SK_f &= B_K \\ -K_c &= C_K S. \end{aligned}$$

Thus, S solves the following equation:

$$S(A + BJ_m D_K C) + SBJ_m C_K S - B_K J_p C - (A_K + B_K DJ_m C_K)S = 0. \quad [2.8]$$

This equation is exactly the same as the Riccati equation [1.18] in T. Thus, *if T (or S) does not exist, then the CSF for given $\overline{G_0}(s)$ and $K_0(s) - D_K$ (or $G_0(s)$ and $K_0(s)$) does not exist.*

REMARK 2.1.– *Proposition 2.3 highlights that the controller $\widehat{K}(s)$ provided by H_2 or H_∞ design on P_{CSF} is non-minimal. It can be shown that the $n - n_K$ non-minimal dynamics in $\widehat{K}(s)$ are assigned to the eigenvalues of $\varphi(X)$ (equation [2.6]) and thus can be assigned by a suitable choice of X (see example in section 2.5).*

Thus, the unique optimal H_∞ controller is non-minimal but is equivalent to the initial controller $K_0(s)$ from the input–output behavior. Note that K_0 can be recovered from the full-order H_2 controller $\widehat{K_{H_2}}$ or the full-order H_∞ controller $\widehat{K_{H_\infty}}$ computing a minimal realization (Matlab® function `minreal`). Since the reduced-order controller K_0 is the global optimal of the CSF H_∞ problem, it can be directly designed using fixed-structure H_∞ synthesis (`hinfstruct`, for instance). This will be illustrated in section 2.5.

2.4. Augmented-order controller case ($n_K > n$)

In the case $n_K > n$, the CSF is directly defined from the three parameters K_c, K_f and $Q(s)$ of the observer-based realization of $K_0(s)$ (see Figure 2.3 and see [ALA 04] for the proof). These parameters can be computed using the procedure presented in section 1.6.

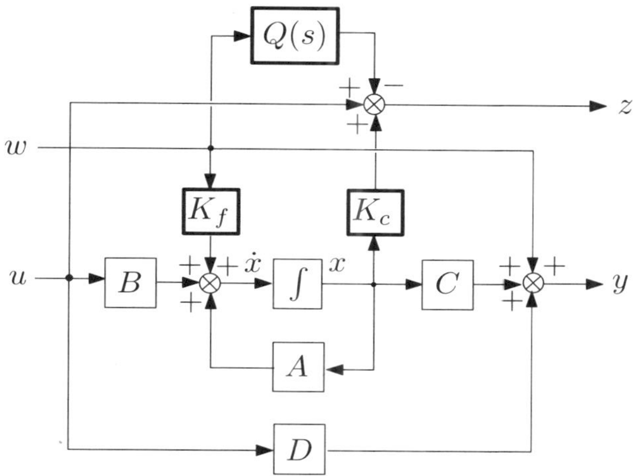

Figure 2.3. *Block diagram of Cross Standard Form P_{CSF} (case $n_K > n$)*

The state-space representation of this CSF reads:

$$P_{CSF}(z) := \left[\begin{array}{cc|cc} A & 0 & K_f & B \\ 0 & A_Q & B_Q & 0 \\ \hline K_c & -C_Q & -D_Q & I_m \\ C & 0 & I_p & D \end{array}\right]. \quad [2.9]$$

2.5. Illustration

2.5.1. *Solving the inverse H_∞-optimal control problem*

The results of this chapter are illustrated on the launcher example $G_0(s)$ presented in section 1.8. Let us consider the system described in equation [1.73] and the initial controller (also used in section 1.3):

$$K_0(s) = \frac{-23s - 32}{s + 12} = -23\frac{s + 1.391}{s + 12} := \left[\begin{array}{c|c} -12 & 4 \\ \hline 61 & -23 \end{array}\right].$$

The only real solution T of [1.18] reads:

$$T = [0.32787 - 0.032787].$$

Let us choose $T^{\#} = T^{+}$, then the CSF (equation [2.2]) reads:

$$P_{CSF}: = \left[\begin{array}{cc|cc} 0 & 1 & 12.079 & 0 \\ 1 & 0 & 21.792 & 1 \\ \hline 3 & 26 & 23 & 1 \\ 1 & 0 & 1 & 0 \end{array}\right].$$

It is easy to check that the optimal H_∞ controller reads:

$$K_\infty(s) = -23\frac{(s+1.391)(s+2.079)}{(s+12)(s+2.079)}.$$

The corresponding Matlab® sequence using function `cor2tfga`[2] is:

```
>> G0=ss([0  1;1  0],[0;1],[1  0],0);
>> K0=ss(-12,4,61,-23);
>> T=cor2tfga(G0,K0);
>> Tml=pinv(T);
>> [a,b,c,d]=ssdata(G0);
>> [AK,BK,CK,DK]=ssdata(K0);
>> CSF=ss(a,[Tml*BK-b*DK b],[-CK*T-DK*c;c],...
   [-DK+DK*d*DK, eye(size(d,2))-DK*d;eye(size(d,1))-d*DK d]);
>> K=hinfsyn(CSF,1,1);
>> zpk(K)

Zero/pole/gain:
-23 (s+1.391) (s+2.079)
-----------------------
   (s+12) (s+2.079)
```

Furthermore, equation [2.6] reads:

$$\varphi(X) = -2.0792 - 0.39801\,X \quad \text{and} \quad \varphi(246.02294) = -100.$$

2 See `http://personnel.isae.fr/daniel-alazard/matlab-packages`

Then, the choice:

$$T^{\#} = T^{+} + 246.0229\,T^{\perp} = [27.5 \quad 244.5]^{\mathrm{T}}$$

leads to a new P_{CSF} and a new optimal H_∞ controller:

$$K_\infty(s) = -23\frac{(s+1.391)(s+100)}{(s+12)(s+100)}.$$

From the previous Matlab® sequence, the corresponding command lines read:

```
>> Nt=null(T);
>> AA=Nt'*(a+b*DK*c)'*Nt;
>> BB=Nt'*c'*BK';
>> X=place(AA,BB,[-100]);X=X';
>> Tml=pinv(T)+Nt*X;
>> CSF=ss(a,[Tml*BK-b*DK b],[-CK*T-DK*c;c],...
   [-DK+DK*d*DK, eye(size(d,2))-DK*d;eye(size(d,1))-d*DK d]);
>> K=hinfsyn(CSF,1,1);
>> zpk(K)

Zero/pole/gain:
-23 (s+1.391) (s+100)
-----------------------
   (s+100) (s+12)

>> zpk(minreal(K))

1 state removed.

Zero/pole/gain:
-23 (s+1.391)
--------------
   (s+12)
```

In both designs, K_∞ is not minimal and $K_\infty = K_0$.

Note that the reduced (first) order and minimal controller $K_0(s)$ can be directly recovered using fixed-structure H_∞ synthesis:

```
>> order1=ltiblock.ss('order1',1,1,1);
>> CL=lft(CSF,order1);
>> [CLopt,gam]=hinfstruct(CL);
Final: Peak gain = 3.76e-13, Iterations = 81
>> K=CLopt.Blocks.order1;
>> zpk(K)

Zero/pole/gain:
-23 (s+1.391)
--------------
   (s+12)
```

2.5.2. *Improving K_0 with frequency-domain specification*

In fact, K_0 has been designed to assign the dominant closed-loop eigenvalues to $-1 \pm i$. Indeed:

$$\text{poles of } \frac{1}{1 - K_0(s)G_0(s)} = \{-1 + i, \ -1 - i, \ -10\}.$$

The magnitude of the frequency-domain response of $K_0(s)$ is plotted in Figure 2.4 (black solid line). Now, let us assume the controller must have a second-order roll-off behavior beyond 10 rad/s and must fulfill the low-pass template also depicted in Figure 2.4 (gray patch). Such a specification can be formulated to attenuate launcher flexible modes that are not taken into account in the design model $G_0(s)$.

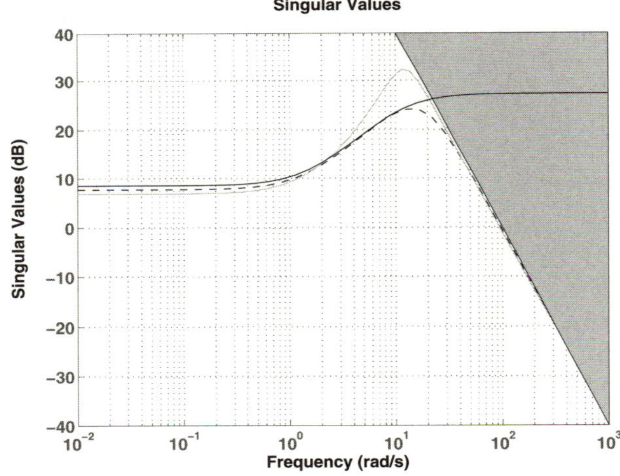

Figure 2.4. *Frequency-domain responses (magnitude) of $K_0(s)$ (black solid line), $K_W(s)$ (dashed line), $K_{A,W}(s)$ (gray solid line) and template (gray patch)*

This specification can be handled, in the H_∞ framework, in weighting the closed-loop transfer from a disturbance on the plant output (measurement noise) to the plant input[3] u. It is obvious that, in the standard problem associated with the CSF (see Figure 2.2), the plant output y is directly linked to the exogenous input w and the

3 Such a transfer reads: $K(I_m - KG)^{-1}$ (with positive feedback).

plant input u is directly linked to the controlled output z. Then, in order to take into account this frequency-domain specification, we can augment this standard problem with a noise w' acting on the measurement y and weighted by a second-order high-pass filter (in order to get a -40 dB/dec roll-off behavior). The augmented CSF $P_{CSF,W}(s)$ is then depicted in Figure 2.5. The high-pass filter $W(s)$ is in fact a second-order derivative filter whose poles $(\frac{-1000}{\sqrt{2}}(1 \pm i))$ are introduced for reasons of properness. The gain g allows the roll-off cut-off frequency to be adjusted and is tuned by a try and error procedure. The tuning $g = 0.02$ provides a fourth-order H_∞-optimal controller $K_W(s)$ whose frequency response is depicted in Figure 2.4 (dashed line). The template is now fulfilled and we can check that the closed-loop dominant dynamics is assigned to the nominal values $-1 \pm i$. Indeed:

$$\text{poles of } \frac{1}{1 - K_W(s)G_0(s)} = \{-1 \pm i,\ -9.7947,\ -14.272 \pm 12.985\,i,\ -4.86\,10^5\}.$$

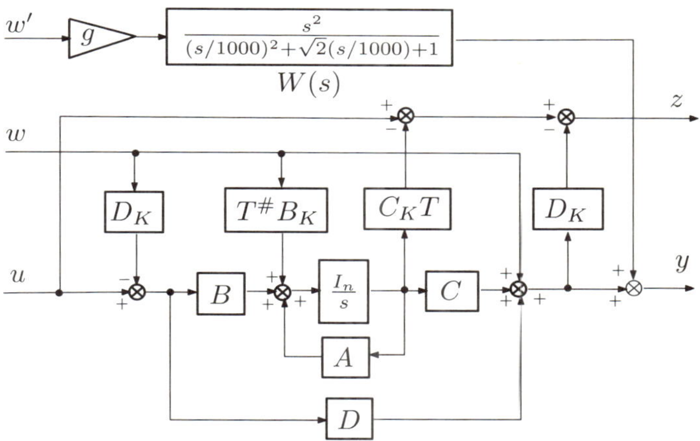

Figure 2.5. *Augmented Cross Standard Form $P_{CSF,W}(s)$ to take into account roll-off specification (with $T^\# = [27.5 \quad 244.5]^T$)*

From the previous Matlab® sequence (see page 63), the corresponding command lines read:

```
>> CSF_W=ss(a,[0*b Tm1*BK-b*DK b],[-CK*T-DK*c;c],...
   [0 -DK+DK*d*DK,eye(size(d,2))-DK*d;1 eye(size(d,1))-d*DK d]);
>> W=tf([1 0 0],[1/1000000 sqrt(2)/1000 1]);
>> [K_W,cl,gam]=hinfsyn(CSF_W*append(0.02*W,1,1),1,1);
>> % Closed loop eigenvalues:
>> damp(feedback(G0,K_W,1))
```

Eigenvalue	Damping	Freq. (rad/s)
−1.00e+00 + 1.00e+00i	7.07e−01	1.41e+00
−1.00e+00 − 1.00e+00i	7.07e−01	1.41e+00
−9.79e+00	1.00e+00	9.79e+00
−1.43e+01 + 1.29e+01i	7.41e−01	1.92e+01
−1.43e+01 − 1.29e+01i	7.41e−01	1.92e+01
−4.86e+05	1.00e+00	4.86e+05

2.5.3. *Improving K_0 with phase lead*

In this section, the objective is to assign the plant main dynamics to the prescribed values $-1 \pm i$ taking into account the actuator dynamics $A(s)$ modeled by a first-order system:

$$A(s) = \frac{5}{s+5} := \left[\begin{array}{c|c} -5 & 5 \\ \hline 1 & 0 \end{array}\right].$$

While it is easy to design a first-order controller $K_0(s)$ to assign the second-order dynamics of the nominal plant model $G_0(s)$, it could be more delicate to obtain the same result on the more representative model of the plant including its avionics $G_A(s) = G_0(s)A(s)$. This is particularly the case when this avionics, here $A(s)$, has a bandwidth (5 rad/s) a little bit too slow with respect to the required closed-loop dynamics ($\sqrt{2}$ rad/s). Indeed, the closed-loop eigenvalues do not any more meet the specifications:

$$\text{poles of } \frac{1}{1 - K_0(s)G_0(s)A(s)} = \{-0.78 \pm 1.59\,i, -2.45, -13\}. \qquad [2.10]$$

The damping ratio of the dominant mode is now lower than 0.5 and the Nichols plot of $-K_0(s)G_0(s)A(s)$, depicted in Figure 2.7, reveals a too weak phase margin (23.9 deg), in comparison with the nominal Nichols plot of $-K_0(s)G_0(s)$. To work around this problem, the controller must include a phase lead. Such a controller can be easily designed, without any additional tuning parameters, using the CSF augmented by the actuator dynamics $A(s)$ on the control input u according to Figure 2.6.

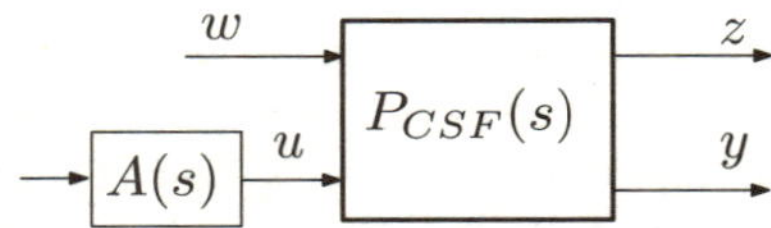

Figure 2.6. *H_∞ standard problem $P_A(s)$: Cross Standard Form P_{CSF} augmented with actuator dynamics $A(s)$*

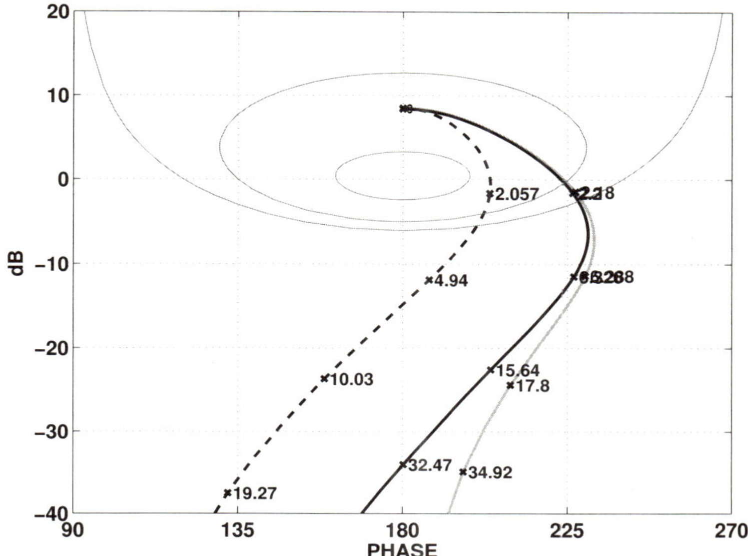

Figure 2.7. *Nichols plots of* $-K_0(s)G_0(s)$ *(gray solid line),* $-K_0(s)G_0(s)A(s)$ *(dashed line) and* $-K_A(s)G_0(s)A(s)$ *(black solid line)*

The full-order H_∞ design on the problem $P_A(s)$ provides a third order controller with a very fast pole ($<-10^5$ rad/s), which can be removed (reduced). The second-order controller $K_A(s)$ thus obtained allows the required assignment for the plant dominant dynamics and a good phase margin (43.4 deg) to be recovered. The phase lead of controller $K_A(s)$ is highlighted on the Nichols plot of $-K_A(s)G_0(s)A(s)$, depicted in Figure 2.7. From the previous Matlab® sequence (see page 63), the corresponding command lines to illustrate such a design are as follows:

```
>> Adyn=tf([5],[1 5]);
>> CSF_A=CSF*append(1,Adyn);
>> [K_A,cl,gam]=hinfsyn(CSF_A,1,1);
>> zpk(K_A)

Zero/pole/gain:
 -403009219.7038 (s+1.382) (s+5)
-------------------------------------
(s+8.64e05) (s+99.74) (s+12.26)

>> K_A=red_fast(K_A,-1000);
>> zpk(K_A)
```

```
Zero/pole/gain:
-466.4811 (s+1.382) (s+5.001)
---------------------------------
    (s+99.74) (s+12.26)

>> damp(feedback(G0*Adyn,K_A,1))

        Eigenvalue                  Damping      Freq. (rad/s)

 -1.00e+00 + 1.00e+00i             7.07e-01        1.41e+00
 -1.00e+00 - 1.00e+00i             7.07e-01        1.41e+00
 -5.00e+00                         1.00e+00        5.00e+00
 -1.00e+01                         1.00e+00        1.00e+01
 -1.00e+02                         1.00e+00        1.00e+02
```

The reduction of the full-order H_∞ controller involves a Matlab® reduction function `red_fast.m` of the library `bib1`, which can be also download at: `http://personnel.isae.fr/daniel-alazard/matlab-packages`.

This function removes, in the modal realization of the controller, all the stable eigenvalues located in the complex plane on the left hand of a prescribed negative real value. The steady-state gain of the controller is preserved.

From the previous design, it can easily be seen that the phase lead introduced in the controller $K_A(s)$ increases significantly the magnitude of the controller high-frequency response (more exactly: $\lim_{\omega\to\infty} |K_a(j\omega)| = 466.48$). A more challenging problem is to assign the plant main dynamics to the prescribed values $-1 \pm i$ taking into account the actuator dynamics $A(s)$ and the roll-off template (depicted in Figure 2.4, gray patch) on the frequency-domain response of the controller.

Such a problem can be solved mixing the previous H_∞ standard problems $P_{CSF,W}$ and $P_A(s)$ into a new H_∞ standard problem $P_{A,W}(s)$ according to Figure 2.8. Note that in $P_{A,W}(s)$, the control signal of the augmented plant model $G_A(s) = G_0(s)A(s)$ is weighted in the controlled output z by the first-order actuator dynamics $A(s)$. Thus, in order to obtain the required -40 dB/dec roll-off behavior, the weight $W(s)$ must be now a third-order high-pass filter. Following the procedure proposed in section 2.5.2, this filter is a third-order derivator with a tunable gain g:

$$W(s) = g\frac{s^3}{(s/1000+1)(s^2/1000^2+s/1000+1)} \quad \text{with: } g = 0.01.$$

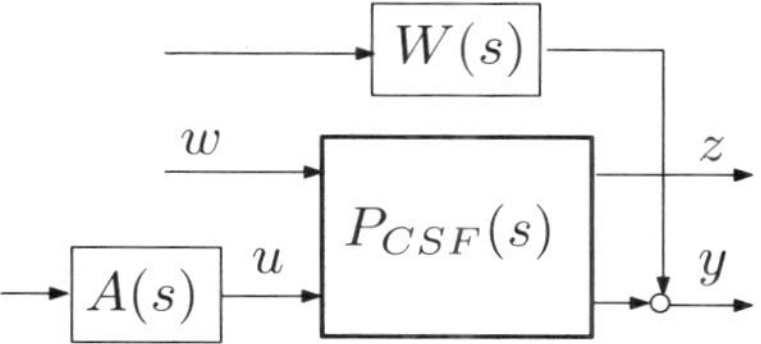

Figure 2.8. *H_∞ standard problem $P_{A,W}(s)$: P_{CSF} with frequency weight $W(s)$ and actuator dynamics $A(s)$*

Then, the full-order H_∞ design on the problem $P_{A,W}(s)$ provides a sixth-order controller with two fast poles (< -1000 rad/s), which can be removed (reduced). The fourth-order controller $K_{A,W}(s)$ thus obtained allows the required assignment for the plant dominant dynamics and a good phase margin (33 deg) to be satisfied. The frequency-domain response of the controller $K_{A,W}(s)$ is plotted in Figure 2.4 (gray solid line): the template is still fulfilled.

From the Matlab® sequence proposed at the end of the section 2.5.2 (see page 65), the command lines corresponding to this design are :

```
>> W=0.01*tf([1 0 0 0],conv([1/1000 1],[1/1000000 1/1000 1]));
>> A=tf([5],[1 5]);
>> CSF_AW=CSF_W*append(W,1,A);
>> [K_AW,cl,gam]=hinfsyn(CSF_AW,1,1);
>> K_AW=red_fast(K_AW,-1000);
>> % Closed loop eigenvalues:
>> damp(feedback(G0*A,K_AW,1))

         Eigenvalue              Damping       Freq. (rad/s)

 -1.00e+00 + 1.00e+00i          7.07e-01        1.41e+00
 -1.00e+00 - 1.00e+00i          7.07e-01        1.41e+00
 -5.00e+00                      1.00e+00        5.00e+00
 -9.98e+00 + 2.07e+00i          9.79e-01        1.02e+01
 -9.98e+00 - 2.07e+00i          9.79e-01        1.02e+01
 -5.70e+00 + 1.06e+01i          4.72e-01        1.21e+01
 -5.70e+00 - 1.06e+01i          4.72e-01        1.21e+01
```

2.6. Pseudo-cross standard form

2.6.1. *A reference model tracking problem*

In this section, the augmented problem $P_{PCSF}(s)$ depicted in Figure 2.9 is considered as a possible candidate CSF for the plant $G_0(s)$ and the initial controller $K_0(s)$.

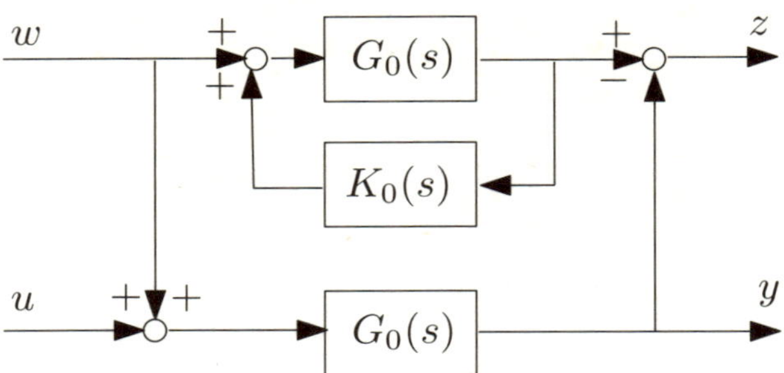

Figure 2.9. *Pseudo-Cross Standard Form:* $P_{PCSF}(s)$

Such a control problem is in fact a reference model tracking problem where the reference model is the nominal closed-loop plant. It is clear that such a standard control problem satisfies the conditions C1, C2 and C3 of the CSF definition 2.3. But there are still some problems with the condition C4:

– such a problem is not a regular H_2 control problem since, for the quite general case of a strictly proper plant G_0, matrices D_{12} and D_{21} (in [2.1]) do not have full-column rank and full-row rank, respectively. Then, the H_2 design will fail on such a standard problem;

– the order of $P_{PCSF}(s)$ is $2n + n_K$ while the order of P_{CSF} proposed in Figures 2.2 or 2.3 is $\max(n, n_K)$. Thus, full-order H_∞ design on $P_{PCSF}(s)$ will provide a $(2n + n_K)$th-order controller. Such a controller will recopy the initial controller $K_0(s)$ in a frequency range (see Figure 2.10) but will require a reduction to remove fast dynamics introduced by the full-order solver and to approximate $K_0(s)$. Of course, a fixed-structure H_∞ synthesis can be used on $P_{PCSF}(s)$, looking for an optimal reduced n_Kth-order controller, to solve the inverse H_∞-optimal control problem. This is illustrated in the following section.

2.6.2. *Illustration*

The following Matlab® sequence considers again the plant $G_0(s) = \frac{1}{s^2-1}$ and the initial controller $K_0(s) = -23\frac{s+1.391}{s+12}$ (see section 2.5) and highlights the results of full-order and reduced-order H_∞ designs on the pseudo-Cross Standard Form $P_{PCSF}(s)$. The frequency-domain responses of the full-order H_∞ controller and the initial controller $K_0(s)$ are plotted in Figure 2.10. The reduction of the full-order H_∞ controller involves the Matlab® reduction function `red_fast.m` of the library `bib1`.

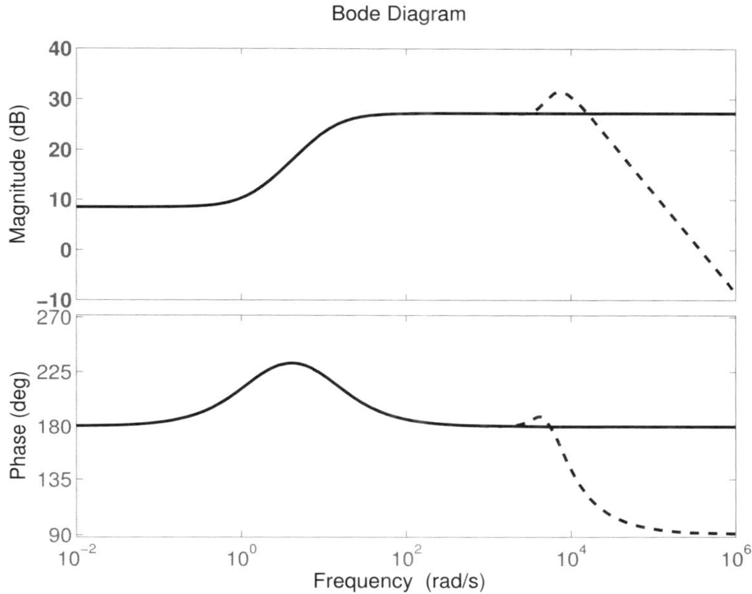

Figure 2.10. *Bode responses of initial controller K_0 (solid line) and full-order H_∞-optimal control on $P_{PCSF}(s)$ (dashed line)*

```
>> G0=ss([0  1;1  0],[0;1],[1  0],0);
>> K0=ss(-12,4,61,-23);
>> CLref=feedback(G0,K0,1);
>> PCSF=[1  -1;0  1]*append(CLref,G0)*[1  0;1  1];
>> % Full-order Hinfty design:
>> [Kinf,CL,GAM]=hinfsyn(PCSF,1,1);
>> damp(Kinf)

        Eigenvalue                Damping        Freq. (rad/s)

   -1.20e+01                      1.00e+00          1.20e+01
   -4.05e+03 + 4.05e+03i          7.07e-01          5.72e+03
   -4.05e+03 - 4.05e+03i          7.07e-01          5.72e+03
   -4.05e+03 + 4.05e+03i          7.07e-01          5.73e+03
   -4.05e+03 - 4.05e+03i          7.07e-01          5.73e+03

>> figure
>> bode(K0)
>> hold on
>> bode(Kinf)
>> Kr=red_fast(Kinf,-1000);
>> zpk(Kr)
```

```
Zero/pole/gain:
-23 (s+1.391)
-------------
   (s+12)

>> % Reduced order design:
>> order1=ltiblock.ss('order1',1,1,1);
>> CL=lft(PCSF,order1);
>> [CLopt,gam]=hinfstruct(CL);

Final: Peak gain = 8.47e-13, Iterations = 60

>> Kred=CLopt.Blocks.order1;
>> zpk(Kred)

Zero/pole/gain:
-23 (s+1.391)
-------------
   (s+12)
```

2.6.3. *Comment*

The main interest of the pure CSF $P_{CSF}(s)$ depicted in Figures 2.2 or 2.3 in comparison with the pseudo-Cross Standard Form of Figure 2.9 is that the plant input u and output y are directly linked to the controlled output z and the exogenous input w, respectively. In addition to regularize the standard H_2 or H_∞ problem, it allows additional specifications, for instance a roll-off specification (as was illustrated in section 2.5.2) to be easily taken into account. That is one of the main objectives of the reverse engineering approach. The reader could try to design a controller meeting the frequency domain template depicted in Figure 2.4 from the pseudo-CSF $P_{PCSF}(s)$: this is not at all an easy task.

2.7. Conclusions

The CSF has been presented here as a particular solution of the inverse optimal control problem. The CSF can be used to mix various synthesis techniques in order to satisfy a multiobjective problem. Indeed, the general idea is to design a first controller to meet some specifications, mainly performance specifications. Then, the CSF is applied on this first solution to initialize a standard problem that will be completed to handle frequency-domain or parametric robustness specifications. This heuristic approach is very interesting when the control law designer wants to:

– take into account a first controller based on *a priori* know-how and physical considerations;

– have access to a modern optimal control framework to manage frequency-domain robustness specifications and the trade-offs between these various specifications.

A multiobjective control design procedure based on the CSF is proposed in [ALA 04] and illustrated on a academic mixed-sensibility (two channels) control problem. Realistic applications of this approach in the field of aeronautics (flight control law design) are described in [ALA 02] and [ALA 06]. In this latter reference, an initial aircraft control law is improved to reduce loads due to turbulence while preserving the flying qualities provided by the initial controller. The CSF for the discrete-time controller and the plant is detailed in section A2.3 (Appendix 2). In [VOI 03], discrete-time CSF is used to meet stability margins taking into account launcher flexible modes from an initial controller designed to meet disturbance rejection specifications.

2.8. Bibliography

[ALA 02] ALAZARD D., "Robust H_2 design for lateral flight control of a highly flexible aircraft", *Journal of Guidance, Control and Dynamics*, vol. 25, no. 3, pp. 502–509, 2002.

[ALA 04] ALAZARD D., VOINOT O., APKARIAN P., "A new approach to multiobjective control design from the viewpoint of the inverse optimal control problem", *Proceedings of the SSSC'04, IFAC Symposium on Systems Structure and Control*, Oaxaca, Mexico, December 2004.

[ALA 06] ALAZARD D., CUMER C., DELMOND F., "Improving flight control laws for load alleviation", *Proceedings of the ASCC2006, 6th Asian Control Conference*, Bali, Indonesia, July 2006.

[APK 06] APKARIAN P., NOLL D., "Nonsmooth H_∞ synthesis", *IEEE Transactions on Automatic Control*, vol. 51, no. 1, pp. 71–86, 2006.

[ARZ 11] ARZELIER D., DEACONU G., GUMUSSOY S., HENRION D., "H2 for HIFOO", *International Conference on Control and Optimization with Industrial Applications*, Ankara, Turkey, August 2011.

[BOY 91] BOYD S., BARRATT G., *Linear Controller Design: Limit of Performance*, Prentice-Hall, 1991.

[DEL 06] DELMOND F., ALAZARD D., CUMER C., "Cross standard form: a solution to improve a given controller with H_2 and H_∞ specifications", *International Journal of Control*, vol. 79, no. 4, pp. 279–287, 2006.

[DOY 89] DOYLE J.C., GLOVER K., KHARGONEKAR P.D., FRANCIS B.A., "State space solutions to standard $\mathcal{H}_2$ and $\mathcal{H}_\infty$ control problem", *IEEE Transactions on Automatic Control*, vol. 34, no. 8, pp. 831–847, 1989.

[FUJ 87] FUJII T., "A new approach to the LQ design from the viewpoint of the inverse regulator problem", *IEEE Transactions on Automatic Control*, vol. 32, no. 11, pp. 995–1004, 1987.

[FUJ 88] FUJII T., KHARGONEKAR P., "Inverse problems in $\mathcal{H}_\infty$ control thery and linear-quadratic differential games", *Proceedings of the 27th Conference on Decision and Control*, IEEE, Austin, TX, pp. 26–31, December 1988.

[GAB 10] GABARROU M., ALAZARD D., NOLL D., "Structured flight control law design using non-smooth optimization", *Proceedings of the 18th IFAC Symposium on Automatic Control in Aerospace*, Nara, Japan, September 2010.

[GAB 12] GABARROU M., ALAZARD D., NOLL D., "Design of flight control architecture using a non-convex bundle method", *Mathematics of Control, Signals, and Systems (MCSS)*, to appear.

[GAH 11] GAHINET P., APKARIAN P., "Structured H_∞ synthesis in MATLAB", *Proceedings of the 18th IFAC World Congress*, Milan, Italy, pp. 1435–1440, 28 August–2 September 2011.

[GUM 09] GUMUSSOY S., HENRION D., MILLSTONE M., OVERTON M., "Multiobjective Robust Control with HIFOO 2.0", *Proceedings of the IFAC Symposium on Robust Control Design*, Haïfa, Israel, June 2009.

[HAR 11] HARTLEY E.N., MACIEJOWSKI J.M., "Designing MPC controllers by reverse-engineering existing LTI controllers", *CoRR*, pp. -1–1, 2011. http://arxiv.org/abs/1109.2816.

[KAL 64] KALMAN R., "When is a linear control optimal?", *Transactions of the ASME, Journal of Basic Engineering*, vol. 86, Issue 1,51, pp. 51–60, 1964.

[LEN 88] LENZ K.E., KHARGONEKAR P.P., DOYLE J.C., "When is a controller $\mathcal{H}_\infty$ optimal?", *Mathematics of Control, Signals and Systems*, vol. 1, pp. 107–122, 1988.

[MOL 73] MOLINARI B.P., "The stable regulator problem and its inverse", *IEEE Transactions on Automatic Control*, vol. AC-18, no. 5, pp. 454–459, 1973.

[SEB 01] SEBE N., "A characterization of solutions to the inverse $\mathcal{H}_\infty$ optimal control problem", *Proceedings of the 40th Conference on Decision and Control*, IEEE, Orlando, FL, pp. 273–278, December 2001.

[SHI 97] SHIMOMURA T., FUJII T., "Strictly positive real $\mathcal{H}_2$ controller synthesis from the viewpoint of the inverse problem", *Proceedings of the 36th Conference on Decision and Control*, IEEE, San Diego, CA, pp. 1014–1019, December 1997.

[SUG 87] SUGIMOTO K., YAMAMOTO Y., "New solution to the inverse regulator problem by the polynomial matrix method", *International Journal of Control*, vol. 45–5, pp. 1627–1640, 1987.

[SUG 98] SUGIMOTO K., "Partial pole-placement by LQ regulator: an inverse problem approach", *IEEE Transactions on Automatic Control*, vol. 43, no. 5, pp. 706–708, 1998.

[VOI 03] VOINOT O., ALAZARD D., APKARIAN P., MAUFFREY S., CLÉMENT B., "Launcher attitude control: discrete-time robust design and gain-scheduling", *Control Engineering Practice*, vol. 11, pp. 1243–1252, 2003.

[ZHO 92] ZHOU K., "Comparison between $\mathcal{H}_2$ and $\mathcal{H}_\infty$ controllers", *IEEE Transaction on Automatic Control*, vol. 37, no. 8, pp. 1261–1265, 1992.

[ZHO 96] ZHOU K., DOYLE J.C., GLOVER K., *Robust and Optimal Control*, Prentice Hall, 1996.

3

Reverse Engineering for Mechanical Systems

3.1. Introduction

In Chapter 2, the cross standard form (CSF) was presented as a *reverse engineering* tool to set up a standard H_∞ problem from a given plant and a given initial controller. This problem was then augmented to take into account frequency domain specifications. In the illustration presented in section 2.5.2, the objective was to shape the given controller to satisfy a roll-off behavior. The efficiency of such an approach to manage various type of specifications (for instance, eigenvalue assignment and frequency-domain specifications in section 2.5.2 or quadratic performance index and frequency-domain specification on the launcher problem in [VOI 03]) and to tune trade-off between these specifications is quite attractive.

But some problems are still open: the value of the closed loop H_∞ norm (γ) of the augmented problem has no physical meaning and cannot be used as an indicator of the distance to the objective. In the example presented in section 2.5.2, the value of $\gamma = \|F_l(P_{CSF,W}(s), K_W(s))\|_\infty$ was 188: from this value, it is not possible to state that it is good or bad, or if the corresponding design meets more or less the template presented in Figure 2.4. One reason is that the exogenous signal w of such a problem (see Figure 2.5) is not an actual disturbance signal acting on the plant but a signal acting simultaneously on the input u, the state derivative $\dot{x}$ and the output y of the plant. In the same way, the control output z is a combination of the input u, the state x and the output y.

Another open question is the choice of the right-inverse $T^\#$ in the standard problem (Figure 2.5). The proposition 2.2 allows us to calculate $T^\#$ in order to assign the non-minimal pole/zero pairs to prescribed values (-100 in the illustration presented section 2.5.2) in the H_∞ controller solution of the pure CSF problem (P_{CSF}). But it can be easily verified that the solution of the augmented problem (Figure 2.5) depends on this choice. The only recommendation we can suggest to the reader is to choose $T^{\#}$ such that the non-minimal pole/zero pairs dynamics is far from the closed-loop dynamics in the left-half plane.

In this chapter, a new reverse engineering scheme is proposed to encounter these problems but this new scheme is restricted to H_∞ reverse engineering for a particular class of systems and initial controllers. This is the general class of mechanical systems and the initial controller is a static state-feedback designed to assign the dynamics of each (dof) to a prescribed second order dynamics and to decouple each of them from the others.

3.2. Context

During primary phases of projects, system characteristics and specifications evolve significantly. Analysis and control design methods that can easily be updated when the specifications or characteristics are modified could be very useful engineering tools. In the general case of mechanical systems, performance specifications often consist of desired pulsations, damping ratios and dof decoupling. To meet such specifications, an H_∞ standard problem based on the acceleration sensitivity function is proposed as a basic scheme. This standard problem can be used to analyze or design control laws taking into account new dynamic elements (actuators dynamics, navigation filter, etc.) or additional specifications (roll off).

This new approach is thus an alternative to the CSF presented in Chapter 2 but its importance is that the H_∞ performance index γ is optimal for the specific value $\widehat{\gamma} = 1$ and the distance to this value is a useful indicator of the distance to objectives when additional specifications are added. The drawback of this approach is that the uniqueness of the optimal solution of the basic problem is no longer guaranteed, while the interest of the CSF lies in the fact that the solution of the basic problem is unique and thus it is guaranteed that the current solution when specifications are added is attracted by the initial controller used to build the CSF. Note also that the CSF is very general and can be applied to any kind of initial controller (static, dynamic, full, reduced or augmented-order, output or state feedback).

In section 3.3, the general mechanical system model, specifications and the initial state-feedback controller are presented. Section 3.4 details H_∞ design based on the *acceleration sensitivity function* and its illustration which is based on the academic example used in section 2.5. In section 3.5, the uniqueness of the solution is discussed through an H_2/H_∞ problem where the basic H_∞ problem is augmented with an H_2 channel between the acceleration disturbance and the control signal. Such a channel is commonly introduced to minimized consumption. In section 3.6, the overall methodology is applied to design the lateral flight control law of an aircraft.

3.3. Model, specifications and initial controller

This chapter concerns multi-variable mechanical systems commonly described by the generalized second-order differential equation [JUN 90, MEI 80]:

$$\mathbf{M}\ddot{q} + \mathbf{D}\dot{q} + \mathbf{K}q = \mathbf{F}u \qquad [3.1]$$

where $q \in \mathbb{R}^n$ is the vector of the n dofs, $\mathbf{M}$, $\mathbf{D}$, $\mathbf{K}$ are, respectively, the $n \times n$ mass, damping and stiffness matrices. $u \in \mathbb{R}^m$ is the vector of the m control signals and $\mathbf{F}$ is the $n \times m$ input matrix. It is assumed that each dof is actuated, that is: $m \geq n$ and $\mathbf{F}$ is full row rank (right-invertible).

Considering the state vector $x = \begin{bmatrix} q^{\mathrm{T}} & \dot{q}^{\mathrm{T}} \end{bmatrix}^{\mathrm{T}}$, a state-space realization of [3.1] reads:

$$\dot{x} = \begin{bmatrix} 0 & I_n \\ -\mathbf{M}^{-1}\mathbf{K} & -\mathbf{M}^{-1}\mathbf{D} \end{bmatrix} x + \begin{bmatrix} 0 \\ \mathbf{M}^{-1}\mathbf{F} \end{bmatrix} u = Ax + Bu \qquad [3.2]$$

For such systems, the basic performance specification is often expressed as follows:

– each dof q_i must have a second-order dynamic behavior characterized by a pulsation ω_i and a damping ratio ξ_i;

– dynamic decoupling between dof is required.

For instance, in the field of flight control, flying qualities are expressed in terms of pulsations and damping ratios that are required for the rigid modes (short-term mode, Dutch-roll mode, etc.) and yaw/roll decoupling [ALA 02]. For spacecraft formation flying, decoupling and second-order behavior for each of the three relative attitudes of a spacecraft with respect to one another are commonly specified [GAU 06]. If such specifications do not concern the vector q used to derive equation [3.1] but the vector of controlled variables η, such that $q = T\eta$ where T is regular, then the following developments can be applied to the new second-order system:

$$T^{\mathrm{T}}\mathbf{M}T\ddot{\eta} + T^{\mathrm{T}}\mathbf{D}T\dot{\eta} + T^{\mathrm{T}}\mathbf{K}T\eta = T^{\mathrm{T}}\mathbf{F}u \qquad [3.3]$$

In the earliest phase of development of such a mechanical system, it is also often assumed that each dof position q_i and rate $\dot{q}_i$ are measured. Then, the direct inspection of equation [3.1] allows a control law to be derived:

$$u = \mathbf{F}^{\#}\left([\mathbf{K} - \mathbf{M}\text{diag}(\omega_i^2)]q + [\mathbf{D} - \mathbf{M}\text{diag}(2\xi_i\omega_i)]\dot{q}\right) \quad [3.4]$$

$$= \mathbf{K_0}\, x \text{ with:} \quad [3.5]$$

$$\mathbf{K_0} = \mathbf{F}^{\#}[\mathbf{K} - \mathbf{M}\text{diag}(\omega_i^2)\ \mathbf{D} - \mathbf{M}\text{diag}(2\xi_i\omega_i)] \quad [3.6]$$

$\mathbf{F}^{\#}$ is a right-inverse of $\mathbf{F}$, such that $\mathbf{F}\mathbf{F}^{\#} = I_n$. $\mathbf{F}^{\#}$ is, in fact, the control allocation matrix and describes the way forces and torques (calculated to control q) are distributed on the set of actuators. A simple solution consists of selecting the minimal length solution, that is $\mathbf{F}^{\#} = \mathbf{F}^{+} = \mathbf{F}^{\mathrm{T}}(\mathbf{F}\mathbf{F}^{\mathrm{T}})^{-1}$. This design is based on the general second-order differential system of equation [3.1] and mechanical "know-how". The resulting closed-loop second-order equation reads:

$$\mathbf{M}\left[\ddot{q} + \text{diag}(2\xi_i\omega_i)\dot{q} + \text{diag}(\omega_i^2)q\right] = 0 \quad [3.7]$$

This equation highlights that desired dynamic behavior is met: if a disturbance acts on the dof q_i, then $q_i(t)$ converges toward 0 with the specified dynamics and is decoupled from the other dof (i.e. $q_j(t) = 0,\ \forall j \neq i,\ \forall t$).

During more advanced phases of the project development, it could be interesting to take into account new details and characteristics of the system; for instance: actuators and sensor dynamics, new actuator and sensor configurations, a navigation filter to estimate q_i and $\dot{q}_i$, parametric uncertainties, etc. From the analysis point of view, tools and methods allowing us to analyze whether specifications are still met when these new elements are considered might be useful. From the synthesis point of view, tools and methods to take into account new components and/or additional specifications to improve the control design are also required.

Control law design methods (Linear Quadratic (LQ), eigenstructure assignment, H_∞ design, etc.) can be used to develop such tools in a more or less suitable way. In any case the first step to be performed is to recover the basic control law proposed in equation [3.4] using a candidate approach. Eigenstructure assignment is a very useful approach to easily handle decoupling [TIS 96, CHA 89, MAG 02]. LQ design using the implicit model reference approach [HIC 96] also allows such kinds of specifications to be handled. In Appendix A3.1 and A3.2, it is shown how the control law [3.4] can be recovered using these approaches. These approaches are restricted to state-feedback design and are not suitable to handle either new frequency domain specifications or a new model augmented with actuator dynamics. Extending these approaches to static or dynamic output feedback leads to the introduction of additional tuning parameters.

Modern optimal control techniques [ZHO 96] (H_∞ synthesis, μ-synthesis, μ-analysis, mixed H_2/H_∞, etc.) based on the well-known augmented standard problem offer a very interesting general framework: the block diagram description of the standard problem allows us to easily include new dynamics. This standard problem can be augmented by the Linear Fractional Transformation (LFT) representation of parametric uncertainties for analysis purpose (μ-analysis) or synthesis purpose (μ-synthesis). It can also be augmented by an LFT representation of measured varying parameters to perform Linear Parameter Varying (LPV) design [APK 95]. Singular values (or structured singular value) of the frequency response of $F_l(P, K)$ provide a sufficient condition for the frequency domain specifications to be reached or indicate how far a given controller is from these specifications. Classical H_∞ approaches such as Mixed-Sensitivity or Loop Shaping can very easily handle frequency domain specifications but are not appropriate to handle dynamic decoupling [SUZ 06]. These approaches do not allow us to recover the solution of equation [3.4] without the use of complex frequency-domain weighting matrices with non-null cross-coupling terms. The standard problem proposed in section 3.4 allows us to obtain this nominal solution and involves very simple frequency weighting functions directly defined by the specifications in terms of pulsation and damping ratio for each dof.

3.4. H$_\infty$ design based on the acceleration sensitivity function

3.4.1. *General results*

The proposed H_∞ design in fact weights the acceleration sensitivity function according to Figure 3.2. The disturbance w acts on the acceleration vector $\ddot{q}$ and the controlled output z weighs the acceleration with the diagonal weighting matrix:

$$W_q = \begin{bmatrix} \frac{s^2+2\xi_1\omega_1+\omega_1^2}{s^2} & 0 & \cdots & & \cdots & 0 \\ 0 & \ddots & 0 & & & \vdots \\ \vdots & 0 & \frac{s^2+2\xi_i\omega_i+\omega_i^2}{s^2} & 0 & & \vdots \\ \vdots & & 0 & \ddots & & 0 \\ 0 & \cdots & \cdots & 0 & & \frac{s^2+2\xi_n\omega_n+\omega_n^2}{s^2} \end{bmatrix} \qquad [3.8]$$

W_q is the inverse of the desired acceleration sensitivity function obtained with the system of equation [3.1] and the control law of equation [3.4]. For a dof i, the desired acceleration sensitivity function is depicted in Figure 3.1 and specifies that low-frequency disturbances on the acceleration must be perfectly rejected while high-frequency disturbances (beyond the bandwidth ω_i of this dof) are not at all rejected. Note also that in the field of aerospace vehicle control, most of the perturbations are expressed as disturbing forces or torques acting on the input of the plant. Thus, W_q can also be defined as a function of the magnitude of the disturbance, the rejection ratio required on the position q and the gain $\mathbf{M}^{-1}$.

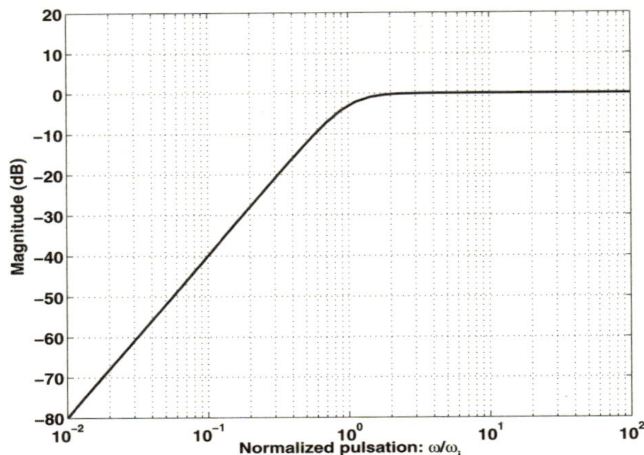

Figure 3.1. *Desired disturbance rejection profile on the acceleration* W_q^{-1}

The multi-variable weight W_q is diagonal and depends only on the specifications $\omega_i,\ \xi_i; i = 1...n$ and not on the system dynamic parameters **M**, **K**, **D** and **F**. It can be easily shown [FEZ 08] that classical input or output sensitivity functions, commonly used by the mixed sensitivity approach, are non-diagonal and depend on the system dynamic parameters.

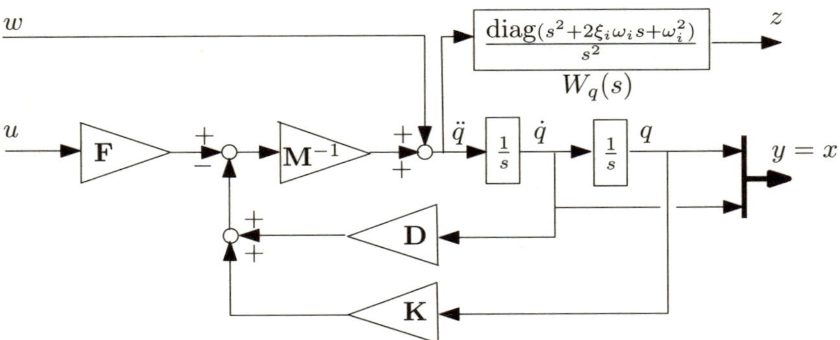

Figure 3.2. H_∞ *standard problem* $P_{ASF}(s)$ *weighting the acceleration sensibility function*

The standard problem $P_{ASF}(s)$ presented in Figure 3.2 is not minimal. H_∞ solvers fail on such a problem because $T_{w\to z}$ has inobservable eigenvalues on the imaginary axis (more particularly at 0). To overcome this difficulty, the synthesis is performed on the minimal realization shown in Figure 3.3. Thus, a minimal state space representation of the standard problem reads:

$$P_{ASF}(s) = \left[\begin{array}{cc|c|c} 0_n & I_n & 0_n & 0_{n\times m} \\ -\mathbf{M}^{-1}\mathbf{K} & -\mathbf{M}^{-1}\mathbf{D} & I_n & \mathbf{M}^{-1}\mathbf{F} \\ \hline \mathbf{M}\text{diag}(\omega_i^2) - \mathbf{K} & \mathbf{M}\text{diag}(2\xi_i\omega_i) - \mathbf{D} & I_n & \mathbf{F} \\ \hline I_n & 0_n & 0_n & 0_{n\times m} \\ 0_n & I_n & 0_n & 0_n \end{array}\right] \quad [3.9]$$

and shows a unitary direct feedthrough between w and z. Therefore,

$$\forall K,\ \|F_l(P_{ASF}, K)\|_\infty \geq 1$$

but the nominal control law of equation [3.6] is a (global) optimal solution for this problem: $\|F_l(P_{ASF}, \mathbf{K_0}))\|_\infty = 1$.

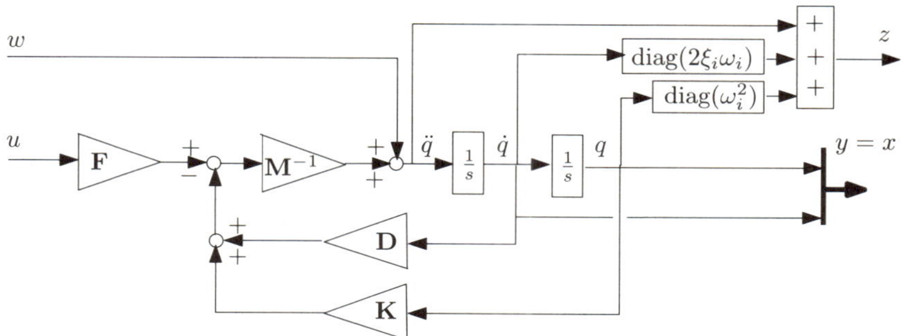

Figure 3.3. *Minimal representation of the H_∞ standard problem $P_{ASF}(s)$ presented in Figure 3.2*

Indeed, the state space representation of $F_l(P_{ASF}, \mathbf{K_0})$ reads:

$$F_l(P_{ASF}, \mathbf{K_0}) = \left[\begin{array}{cc|c} 0_n & I_n & 0_n \\ -\text{diag}(\omega_i^2) & -\text{diag}(2\xi_i\omega_i) & I_n \\ \hline 0_n & 0_n & I_n \end{array}\right] = I_n \quad [3.10]$$

In general, the solution to the H_∞ control problem is not unique and there is no possibility to ensure that the solution K_∞ provided by a H_∞ solver will coincide with $\mathbf{K_0}$. It means that the frequency response of $F_l(P_{ASF}, K_\infty)$ can be below 1 (0 dB) for some frequencies. From a practical point of view, H_∞ solvers provide a controller whose order is equal to the standard problem order (i.e. $2n$). In fact, the poles of this controller are very fast and this controller can be reduced to its DC gain. We can also use fixed-structure H_∞ synthesis to specify a static controller in the design process.

3.4.2. *Illustration*

Let us consider one more time the academic example presented in sections 2.5 and 1.8 where:

$$G_0(s) = \frac{1}{s^2 - 1} \equiv \left[\begin{array}{cc|c} 0 & 1 & 0 \\ 1 & 0 & 1 \\ \hline 1 & 0 & 0 \end{array}\right] \qquad [3.11]$$

The controller $K_0(s) = \frac{-23s-32}{s+12}$ was in fact designed to assign the launcher dynamics to $-1 \pm i$, that is the closed-loop dominant dynamics corresponds to a second-order with a pulsation $\omega = \sqrt{2}\,(rad/s)$ and a damping ratio $\xi = \sqrt{2}/2$. Thus, the desired acceleration sensitivity function reads:

$$W_q^{-1}(s) = \frac{s^2}{s^2 + 2s + 2}$$

and the corresponding H_∞ standard problem $P_0(s)$ weighting the acceleration sensitivity function is depicted in Figure 3.4 (SIMULINK file format).

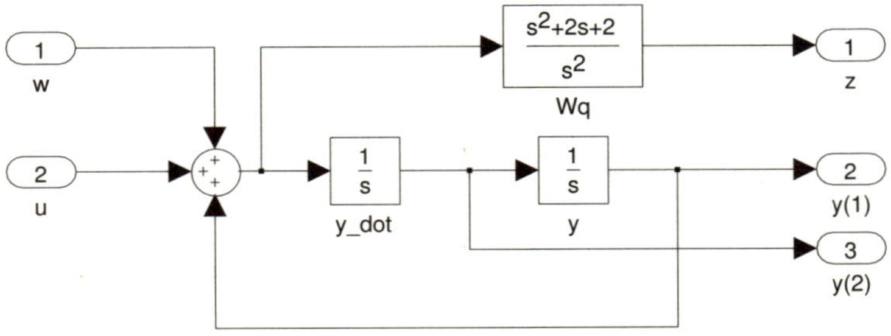

Figure 3.4. *H_∞ standard problem $P_0(s)$ (weighting the acceleration sensibility function) associated with plant $G_0(s)$ and controller $K_0(s)$ (file `std_asf_1.mdl`)*

The direct application of equation [3.6] gives:

$$\mathbf{K_0} = [-3 \quad -2]$$

Such a static output (or state) feedback allows the template on the closed-loop acceleration sensitivity function to be met and saturated at any frequency ω. Indeed:

$$|F_l(P_0(j\omega), [-3 \quad -2])| = 1 \quad \forall\, \omega$$

The following Matlab® sequence shows the solutions provided by various H_∞ solvers:

```
>> [a,b,c,d]=linmod('std_asf_1');
>> std=minreal(ss(a,b,c,d));
2 states removed.
>> K0=[-3 -2];
>> CL_0=lft(std,K0);
>> figure
>> sigma(CL_0)     % ==> the template is saturated for all frequencies !
>> hold on
>> % Riccati based Hinf optimal controller:
>> [Kinf,CL_1,gam]=hinfsyn(std,2,1);
>> K_ric=dcgain(Kinf)      % Controller reduction

K_ric =

   -3.0000   -4.9999

>> CL_ric=lft(std,K_ric);
>> norm(CL_ric,'inf')      % Reduction is OK !

ans =

    1.0000

>> sigma(CL_ric); % ==> in a frequency-domain band; the closed-loop
>> % control system is more performant than the specification!
>> % LMI based Hinf optimal controller:
>> [Kinf,CL_1,gam]=hinfsyn(std,2,1,'METHOD','lmi');
>> K_lmi=dcgain(Kinf)      % Controller reduction: very close to K0

K_lmi =

   -3.0019   -2.0055

>> CL_lmi=lft(std,K_lmi);
>> norm(CL_lmi,'inf')      % Reduction is OK !

ans =

     1

>> sigma(CL_lmi);  % ==> the specification is saturated for all
>> % frequencies
>> % Structured static control design with HINFSTRUCT:
```

```
>> Kstat=ltiblock.gain('Kstat',1,2);
>> CL_K=lft(std,Kstat);
>> [CL,gam]=hinfstruct(CL_K);
Final: Peak gain = 1, Iterations = 1
>> K_struct=ss(CL.Blocks.Kstat)

d =
       u1   u2
   y1  -3   -3

Static gain.
>> sigma(ss(CL)); % ==> in a frequency-domain band; the closed-loop
>> % system is more performant than the specification!
>> % HINSTRUCT with a hand-made initialization:
>> Kstat=ltiblock.gain('Kstat',[-0.01 -0.01]);
>> CL_K=lft(std,Kstat);
>> [CL,gam]=hinfstruct(CL_K);
Final: Peak gain = 1, Iterations = 3
>> K_struct=ss(CL.Blocks.Kstat) % ==> quite the same than previous one.

d =
           u1       u2
   y1   -3.03   -2.984

Static gain.
>> % ==> Thus, reverse engineering is quite interesting to initialize
>> % correctly the controller in HINFSTRUCT:
>> Kstat=ltiblock.gain('Kstat',K0);
>> CL_K=lft(std,Kstat);
>> [CL,gam]=hinfstruct(CL_K);
Final: Peak gain = 1, Iterations = 1
>> K_struct=ss(CL.Blocks.Kstat) % ==> K0 is recovered !

d =
       u1   u2
   y1  -3   -2

Static gain.
>> % A good trick to saturate the template for all frequencies
>> % (and then recover K0) is to minimize the maximum of the Hinfty
>> % norms of CL_K and 1/CL_K (that works for any initilizations
>> % except [0 0]):
>> Kstat=ltiblock.gain('Kstat',[-0.01 -0.01]);
>> CL_K=lft(std,Kstat);
>> [CL,gam]=hinfstruct(blkdiag(CL_K,1/CL_K))
Final: Peak gain = 1, Iterations = 19
>> K_struct=ss(CL.Blocks.Kstat) % ==> K0 is recovered !

d =
       u1   u2
   y1  -3   -2

Static gain.
```

This sequence leads to the following conclusions:

– the reduction of full-order H_∞ controllers (provided by Riccati-based or Linear Matrix Inequalities (LMI)-based solvers) to their DC gains does not affect the closed-loop performance;

– the static controller $K_{\text{Ric}} = \widehat{K}_{\text{Ric}}(0)$ provided by the Riccati-based solver does not saturate the template in a frequency band (see Figure 3.5). That is, in this frequency band, the closed-loop disturbance rejection is better than the specification;

– the static controller $\widehat{K}_{\text{LMI}}(0)$ provided by the LMI-based solver is very closed to $\mathbf{K}_0$ and so saturates the template at any frequency. Therefore, it is possible to recover the nominal static state-feedback $\mathbf{K}_0$ from the DC gain of the full-order controller provided by the LMI solver proposed in the Matlab® function `hinfsyn`, that is detailed in a more general way in Appendix A3.3

– the static controller K_{struct} provided by the structured H_∞ controller solver, when no initialization is provided, does not converge to $\mathbf{K}_0$ and so does not saturate the template. If the design is initialized with $\mathbf{K}_0$, then $K_{\text{struct}} = \mathbf{K}_0$. This shows that $\mathbf{K}_0$ is at least a local minimum for the problem $P_0(s)$. From a methodological point of view, it is worth to mentioning that structured H_∞ controller solvers are very efficient (that will be shown in the next illustrations of this chapter) but the initialization is a key element that cannot be done randomly. In that sense, *reverse engineering* can be very useful to provide a well-conditioned initialization;

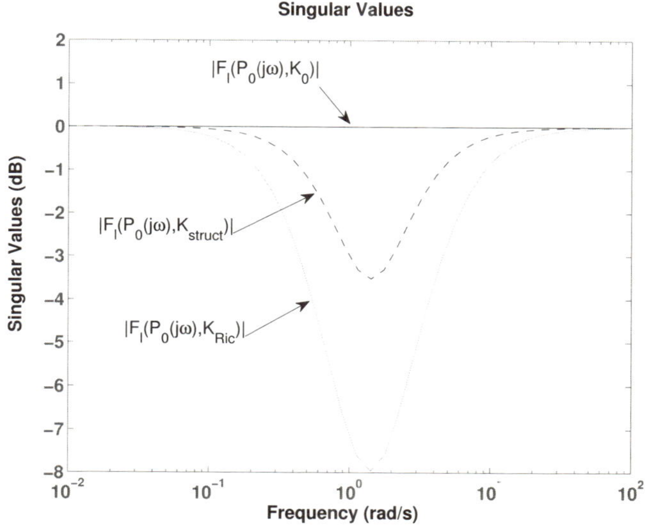

Figure 3.5. *Singular value responses of $F_l(P_0(s), K_i)$ for the various static controller designs*

– a good alternative to recover $\mathbf{K_0}$, from an arbitrary initialization, consists of finding the static controller $\widehat{K}_{\text{struct}}$ such that:

$$\widehat{K}_{\text{struct}} = \arg\min_{K} \max(\|F_l(P_0(s), K)\|_\infty, \|F_l(P_0(s), K)^{-1}\|_\infty)$$

3.4.3. *Analysis on an augmented model*

The H_∞ standard problem weighting the acceleration sensitivity function can be augmented with actuator dynamics $A(s)$, sensors dynamics and navigation filter $N(s)$ (see Figure 3.6) to analyze the degradation of performance in terms of acceleration disturbances rejection of the nominal (initial) controller.

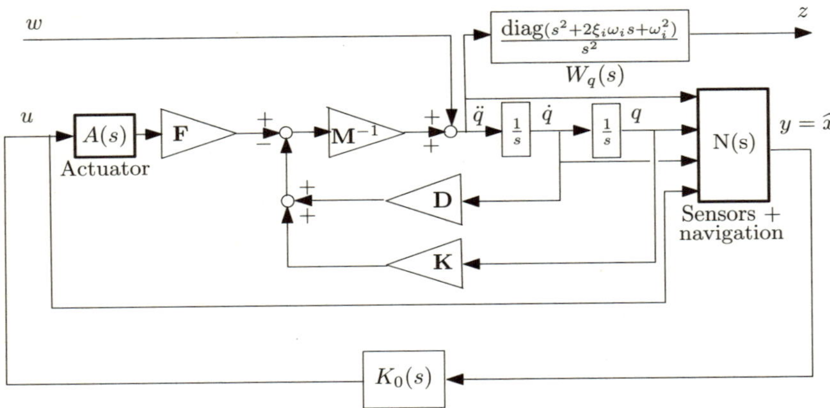

Figure 3.6. *Analysis of the initial controller $K_0(s)$ on a more representative model*

3.4.4. *Illustration*

Following the illustration proposed in section 3.4.2, let us consider the standard problems $P_1(s)$ and $P_2(s)$ depicted in Figures 3.7 and 3.8. $P_1(s)$ is the problem $P_0(s)$ based on the acceleration sensitivity function (see Figure 3.4) augmented with the actuator dynamics $A(s) = \frac{5}{s+5}$ on the control signal u. $P_2(s)$ is like $P_1(s)$ but only the position q is measured. The objective is to analyze the performance of the initial controller $K_0(s) = \frac{-23s-32}{s+12}$ when actuator dynamics $A(s)$ is taken into account.

Table 3.1 highlights the degradation of the performance index γ:

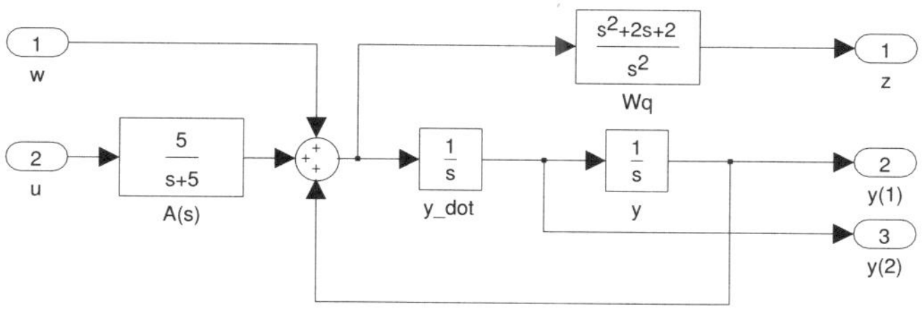

Figure 3.7. *H_∞ standard problem $P_1(s)$ (file* `std_asf_2.mdl`*)*

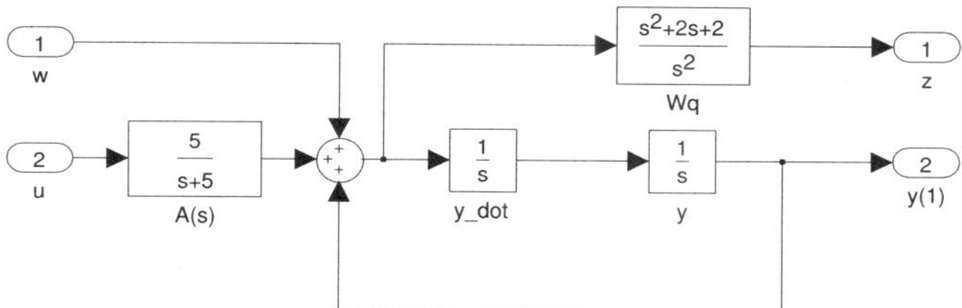

Figure 3.8. *H_∞ standard problem $P_2(s)$ (file* `std_asf_3.mdl`*)*

– γ increases from 1 to 1.2 (which is quite acceptable) when the dynamic position feedback $K_0(s)$ is considered instead of the static state-feedback $\mathbf{K_0}$ on the nominal standard problem $P_0(s)$;

– γ increases up to 1.7 and 2.2 (which is not acceptable) for the static feedback $\mathbf{K_0}$ and the dynamic position feedback $K_0(0)$ cases, respectively, when the actuator dynamics $A(s)$ is taken into account. This degradation can also be interpreted on the closed-loop dynamics that do not meet any more the required assignment (see section 2.5.3, equation [2.10]).

$\|F_l(P_0(s), \mathbf{K_0})\|_\infty = 1$
$\|F_l(P_0(s), [K_0(s)\ 0])\|_\infty = 1.2$
$\|F_l(P_1(s), \mathbf{K_0})\|_\infty = 1.7$
$\|F_l(P_2(s), K_0(s)\|_\infty = 2.2$

Table 3.1. *Performance indexes (γ)*

3.4.5. *Synthesis on an augmented model*

Of course, it is possible to submit to a H_∞ solver the augmented standard problem to find a state or output feedback dynamic controller. This new controller is expected to provide a better performance index γ, that is a better disturbance rejection. More generally, one can consider standard problems depicted:

– in Figure 3.9 where the objective is to take into account actuators dynamics $A(s)$ in the design of the state-feedback;

– in Figure 3.10 where it also assumed that only the position q is measured (output feedback design).

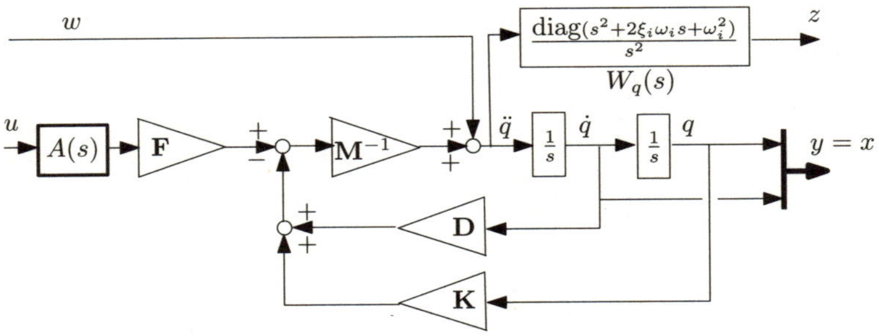

Figure 3.9. *Augmented plant with actuators dynamics*

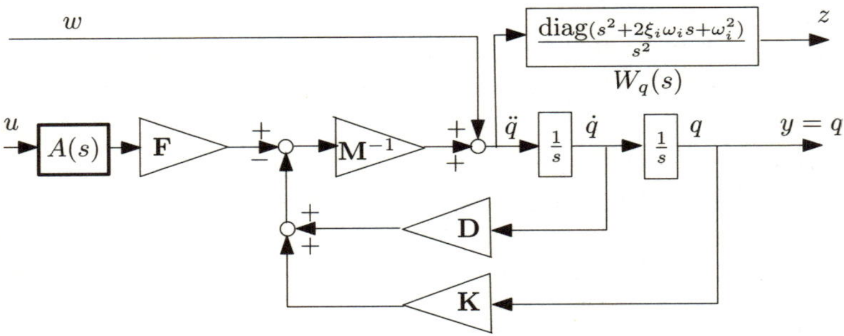

Figure 3.10. *Augmented plant with actuators dynamics for output (position) feedback control design*

Output feedback will refer to the case where only q is measured and state-feedback where q and $\dot{q}$ are measured. Output feedback controllers can be designed using the same methodology on the augmented plant of Figure 3.10. Note that in the output feedback case and/or when actuator dynamics is taken into account, optimal controllers become dynamic and cannot be reduced to static gains anymore.

One of the main interests of the standard problem based on the acceleration sensitivity function is to have a weighting system $W_q(s)$ that is independent of the used measures (and used control inputs). Most of H_∞ standard problems are based on the plant input–output behavior (mixed sensitivity, loop shaping) and they have a weighting system that depends on the used measures. In other words, taking into account that a new measure is available requires redefining the weighting system. This is no longer the case with the acceleration sensitivity-based standard problem. From a methodological point of view, it simplifies consequently the design or the update of the controller to a new measurement configuration which is discussed in the following illustration.

Regarding the interest of a recent fixed-structure H_∞ synthesis w.r.t. to classical full-order synthesis, it will be also highlighted that:

– it is better, from the performance index point of view, to design directly a reduced order controller than to reduce a full-order controller where very fast dynamics commonly appear and can raise implementation problems. This assumes the designer is able to specify the correct order (or even better, the correct structure) of the controller. In the academic illustrations proposed in this chapter, such a task is easy but could be more delicate for some applications;

– the initialization of the controller in structured H_∞ synthesis is a very determinant success factor. In that sense, *a priori* knowledge of an initial controller K_0 can be very useful for the initialization. This is also one of the main concerns of the *reverse engineering* approach developed in this chapter.

Although these two recommendations are illustrated with academic examples, they are quite complex to be kept in mind by the designer faced with more realistic applications.

3.4.6. *Illustration*

The standard problems $P_1(s)$ and $P_2(s)$ depicted in Figures 3.7 and 3.8 are submitted to the various H_∞ design solvers in the following Matlab® sequences.

3.4.6.1. *Sequence on the standard problem $P_1(s)$*

```
>> [a,b,c,d]=linmod('std_asf_2');
>> std=minreal(ss(a,b,c,d));

2 states removed.
>> [Kinf,CL_1,gam]=hinfsyn(std,2,1);
>> gam

gam =
```

```
    1.0010

>> damp(Kinf)

 Eigenvalue        Damping         Freq. (rad/s)

 -1.00e+00         1.00e+00          1.00e+00
 -1.82e+05         1.00e+00          1.82e+05
 -9.06e+05         1.00e+00          9.06e+05

>> Kinfr=red_fast(Kinf,-1e5);
>> norm(lft(std,Kinfr),'inf')
ans =

    1.4813
```

Comments:

– the full-order H_∞ design provides a third-order controller with a very good performance index;

– this controller shows two very fast poles ($< -10^5\ (rad/s)$). The reduction of these poles degrades significantly the performance index.

```
>> order1=ltiblock.ss('order1',1,1,2);
>> CL=lft(std,order1);
>> [CLopt,gam]=hinfstruct(CL);
Final: Peak gain = 1, Iterations = 61
```

Comment:

– the first-order H_∞ design provides directly a reduced order controller with a very good performance index (better than the performance index of the full-order design that is assumed to provide the global optimum; such a result is due to numerical aspects).

3.4.6.2. *Sequence on the standard problem* $P_2(s)$

```
>> [a,b,c,d]=linmod('std_asf_3');
>> std=minreal(ss(a,b,c,d));
2 states removed.
>> [Kinf,CL_1,gam]=hinfsyn(std,1,1);
>> gam

gam =

    1.0004

>> damp(Kinf)
```

```
        Eigenvalue                   Damping        Freq. (rad/s)

  -1.07e+04 + 1.03e+04i          7.20e-01          1.49e+04
  -1.07e+04 - 1.03e+04i          7.20e-01          1.49e+04
  -7.57e+08                      1.00e+00          7.57e+08

>> Kinfr=red_fast(Kinf,-1e5);
>> norm(lft(std,Kinfr),'inf')

ans =

    1.0003
```

Comments:

– in the case of the position output feedback, the reduction of the full-order controller has no consequence on the performance index;

– the weighting filter $W_q(s)$ works in the state-feedback or output feedback cases; its tuning is independent of the measurement vector;

– due to the required phase lead, and the pseudo-derivation of the position to have an estimate of the velocity, the controller order on such a problem must be at least equal to two. This will be used in the next sequence to specify the order for the fixed-structure controller design.

```
>> order2=ltiblock.ss('order2',2,1,1);
>> CL=lft(std,order2);
>> [CLopt,gam]=hinfstruct(CL);
Final: Peak gain = 1.4, Iterations = 59
       Spectral radius 4.52e+04 is close to bound 5e+04
>> Kinit=K0*tf([1/50 1],[1/50 1]);
>> norm(lft(std,Kinit),'inf')

ans =

    2.2027

>> order2=ltiblock.ss('order2',Kinit);
>> CL=lft(std,order2);
>> [CLopt,gam]=hinfstruct(CL);
Final: Peak gain = 1, Iterations = 101
>> K_struct=ss(CLopt.Blocks.order2);
>> zpk(K_struct)

Zero/pole/gain:
-911627.4437 (s+5.004) (s+1.499)
---------------------------------
     (s^2 + 1735s + 2.281e06)
```

Comments:

– the design of a second-order controller, without any initialization, leads to $\gamma = 1.4$ that is not very good in comparison to full-order design;

– the initialization of the design with $K_0(s)\frac{s+50}{s+50}$, that is $K_0(s)$ (first-order) in series connection with a pole/zero cancellation allows the design to improve very efficiently the performance index, from 2.2 (this value was already calculated in Table 3.1) to the optimal value $\widehat{\gamma} = 1$.

3.4.7. *Taking into account a roll-off specification*

The previous illustration has shown that H_∞ design provides controllers with an important phase lead to optimize the disturbance rejection. Such controllers have also a high-pass behavior with a too high magnitude in high frequencies that could be a drawback from the actuator health point of view.

A controller roll-off behavior can be handled using a weight $W_2(s)$ on the control signal u according to the new augmented standard plant depicted in Figure 3.11. $W_2(s)$ is a high-pass filter weighting u in high frequencies. Gains $\mathbf{F}$ and $\mathbf{M}^{-1}$ are introduced in order so that new controlled output z_2 can be compared with controlled output z_1 (assuming that actuator dynamics $A(s)$ has a unitary DC gain). The way $W_2(s)$ can be tuned is illustrated on the control of a Reusable Launch Vehicle (RLV) during atmospheric re-entry in [FEZ 07].

This way of handling the roll-off specification is in fact a result of the first H_∞ approaches based on full order synthesis. The frequency-domain constraint on the controller open-loop frequency response was indirectly taken into account weighting closed-loop sensitivity functions (the most well-known is certainly $K(I + KG)^{-1}$ in the mixed-sensitivity approach based on the fact that, in the roll-off frequency range, $|K(j\omega)G(j\omega)| << |K(j\omega)|$). Today, fixed-structure H_∞ synthesis allows us to handle several performance indexes associated with several standard problems (models) involving the same controller. Considering a two channels H_∞ standard problem defined by two performance indexes $J_{c1}(K) = \|F_l(P_{c1}(s), K)\|_\infty$ and $J_{c2}(K) = \|F_l(P_{c2}(s), K)\|_\infty$ associated with two elementary standard problems $P_{c1}(s)$ and $P_{c2}(s)$, it is possible to design the controller $\widehat{K}$ stabilizing $F_l(P_{c1}(s), \widehat{K}(s))$ and $F_l(P_{c2}(s), \widehat{K}(s))$ such that:

$$\widehat{K} = \arg\min_K \max(\|F_l(P_{c1}(s), K)\|_\infty, \|F_l(P_{c2}(s), K)\|_\infty)$$

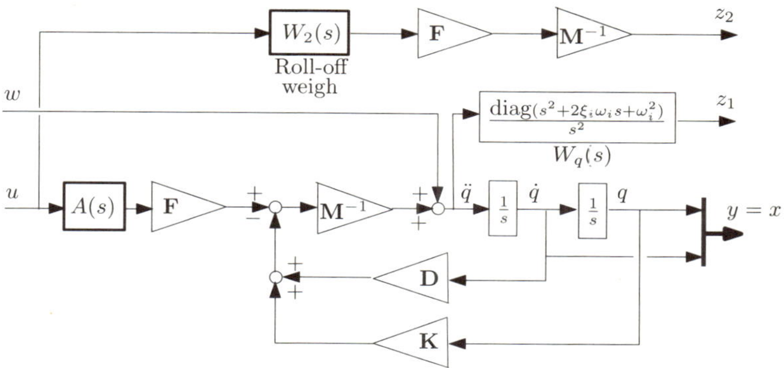

Figure 3.11. *Augmented plant will roll-off criterion based on W_2 high-pass filter*

This specific feature is very interesting to take into account a roll-off specification:

– let us consider an elementary standard problem $P_{c1}(s)$ for performance achievement, that is to say: if $J_{c1}(K) = \|F_l(P_{c1}(s), K)\|_\infty \leq 1$ then, the controller $K(s)$ is efficient. Then, the template $B(s)$ on the open-loop controller frequency response (as it is defined, for instance, in Figure 2.4) can be handled taking into account the second problem $J_{c2}(K) = \|B(s)^{-1}K(s)\|_\infty$. Indeed, if $J_{c2}(K) \leq 1$, then $\sigma_{max}(K(j\omega)) \leq |B(j\omega)| \; \forall \; \omega$, that is to say: the controller $K(s)$ satisfies the template $B(s)$;

– such a two H_∞ channels design guarantees the stability of the controller $K(s)$. *Strong stabilization*, that is the stabilization of the closed-loop plant with a stable controller, is a property commonly required for the implementation purpose [FEI 11].

The two channels are summarized in Figure 3.12. The second channel is just the series connection of the controller with the inverse of the template. Note that this channel can also be expressed using a Linear Fractional Representation (LFR):

$$B(s)^{-1}K(s) = F_l(P_{c2}(s), K(s)) \quad \text{with: } P_{c2}(s) = \begin{bmatrix} 0 & B(s)^{-1} \\ I_m & 0 \end{bmatrix}$$

In such an approach, it is recommended to normalize to one the required performance indexes of the various elementary problems. That is the case when the performance is managed using a standard problem based on an acceleration sensitivity function ($P_{c1}(s) = P_{ASF}(s)$) but it is not the case using the CSF $P_{CSF}(s)$ presented in Chapter 2.

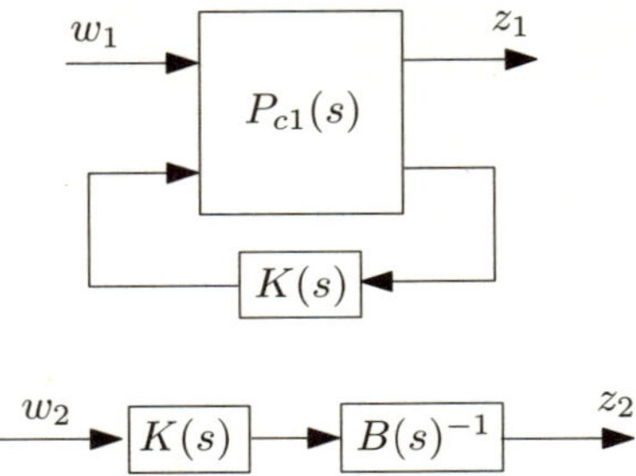

Figure 3.12. *A two channel H_∞ problem to handle strong stabilization with a template $B(s)$ on the open-loop frequency response of the controller $K(s)$*

3.4.8. *Illustration*

The problem $P_2(s)$ depicted in Figure 3.8 is reconsidered together with the frequency domain template depicted in Figure 2.4 to be met by the controller frequency response. The solution obtained using fixed-structure H_∞ synthesis can thus be compared with the solution $K_{A,W}(s)$ obtained in section 2.5.3 using the CSF-based approach.

The template depicted in Figure 2.4 can be bounded by the following low-pass filter:

$$B(s) = 100\frac{s^2/1000^2 + \sqrt{2}s/1000 + 1}{s^2/10^2 + \sqrt{2}s/10 + 1}$$

Then, the following Matlab® sequence shows how the two channels H_∞ problem can be solved, defined by:

– $\|F_l(P_2(s), K(s))\|_\infty \leq 1$ (i.e template on the acceleration sensitivity function to satisfy the disturbance rejection performance),

– $\|B(s)^{-1}K(s)\|_\infty \leq 1$ (i.e template on controller frequency response to satisfy the required roll-off behavior).

```
>> [a,b,c,d]=linmod('std_asf_3');
>> std=minreal(ss(a,b,c,d));
2 states removed.
>> Template=100*tf([1/1000^2 sqrt(2)/1000 1],[1/100 sqrt(2)/10 1]);
>> K_4th=ltiblock.ss('order4',4,1,1);
>> pb_1=lft(std,K_4th);
>> pb_2=1/Template*K_4th;
```

```
>> [CL,gam]=hinfstruct(blkdiag(pb_1,pb_2));
Final: Peak gain = 1.42, Iterations = 120
>> K_struct=ss(CL.Blocks.order4);
>> damp(lft(std,K_struct))

        Eigenvalue              Damping       Freq. (rad/s)

  -1.02e+00 + 1.00e+00i        7.12e-01         1.43e+00
  -1.02e+00 - 1.00e+00i        7.12e-01         1.43e+00
  -4.49e+00                    1.00e+00         4.49e+00
  -3.13e+00 + 1.00e+01i        2.99e-01         1.05e+01
  -3.13e+00 - 1.00e+01i        2.99e-01         1.05e+01
  -1.39e+01 + 1.05e+01i        7.99e-01         1.74e+01
  -1.39e+01 - 1.05e+01i        7.99e-01         1.74e+01
```

Comments:

– following the recommendation of the previous illustration and due to the second-order roll-off specification, a fourth order is now required for this problem and specified in the design procedure. Note that no initialization is provided. It can be easily checked that the performance index ($\gamma = 1.42$) does not depend on the initialization in this case. It can be explained by the additional constraint (here the roll-off specification) that reduces the number of local minima in the optimization problem;

– let $K_{struct}(s)$ be the fourth order controller provided by this design. The value $\gamma = 1.42$ of the performance index shows that specifications are not met completely. That can also be interpreted on the frequency-domain responses of $W_q^{-1}(s)\,F_l(P_2(s), K_{struct}(s))$, that is the (unweighted) closed-loop acceleration sensitivity function, and $K_{struct}(s)$ presented in Figure 3.13. The templates $W_q^{-1}(s)$ and $B(s)$ and the responses obtained with the controller $K_{A,W}(s)$ designed in section 2.5.3 using the CSF are also plotted for comparison: from the required roll-off point of view, the controller $K_{A,W}(s)$ provides better results but from the disturbance rejection point of view, K_{struct} is better. This value ($\gamma = 1.42$) indicates also that the trade-off between the phase lead and roll-off behaviors in the controller to be designed is very delicate.

– note also that the dominant closed-loop dynamics is correctly assigned with K_{struct}.

The main interest of the acceleration sensitivity function approach is to set up the H_∞ problem directly from the specifications (the pulsation ω and the damping ratio ξ of the dominant dynamics, the template specified on the frequency response of the controller) without any other tuning parameters.

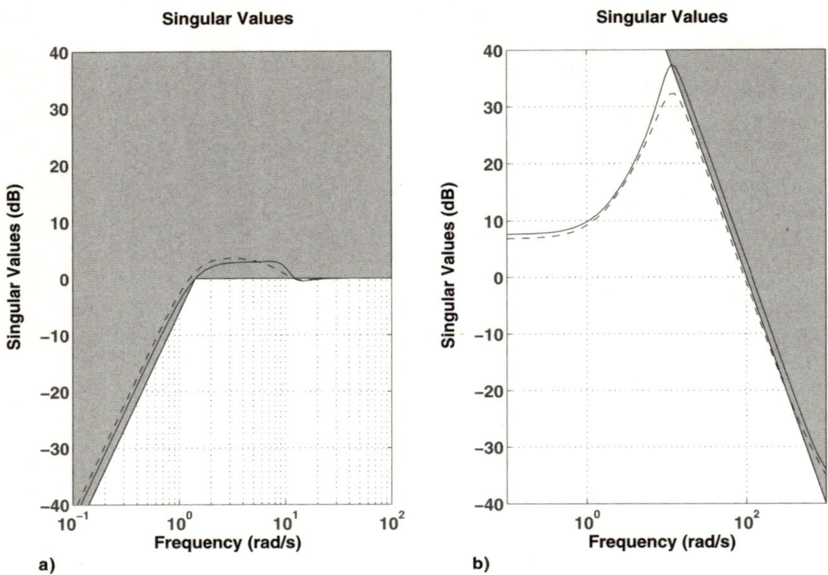

Figure 3.13. *a) Frequency-domain responses of acceleration sensitivity functions* $W_q^{-1}(s)\, F_l(P_2(s), K_{struct}(s))$ *(solid) and* $W_q^{-1}(s)\, F_l(P_2(s), K_{A,W}(s))$ *(dashed) and template* $W_q^{-1}(s)$ *(patch). b) Frequency-domain responses of* $K_{struct}(s)$ *(solid) and* $K_{A,W}(s)$ *(dashed) and template* $B(s)$ *(patch)*

3.4.9. *Taking into account an integral term*

Although it ensures good closed-loop disturbance rejection, the previous methodology does not ensure: (1) a robust reference input tracking or (2) a perfect rejection on the position q of an acceleration disturbance step w. When a null steady-state error on the response of the position q to a step reference input q_{ref} is required, a very well-known solution consists of adding an integral action in the control law when the open-loop plant has no integral term. Such a solution ensures a null steady error in spite of uncertainties on the plant parameters ($\mathbf{F}$, $\mathbf{M}$, $\mathbf{D}$, $\mathbf{K}$) as long as these uncertainties do not destabilize the closed-loop system. When a perfect rejection on the position q of an acceleration disturbance step w is required, the controller must have at least one integrator, independently of the plant properties (except a plant with a derivative behavior). That is why the problem (2) is more general than the problem (1) and why it is recommended to specify the integral action of the controller as a disturbance rejection problem. Note also that such requirements cannot be directly taken into account in classical mixed-sensitivity or loop shaping H_∞ approaches where a regularization is required to set up a detectable and/or stabilizable standard problem. This regularization often consists of changing the integral weight $1/s$ by a pseudo-integral weight $1/(s+\epsilon)$ and is not completely satisfactory from a methodological point of view since the resulting controller depends on the regularization parameter ϵ and finally does not exhibit a pure integral action. To overcome this problem, it is recommended to impose the integrator in the controller structure.

With these considerations in mind, the basic control problem depicted in Figure 3.2 can be augmented with an integral term according to Figure 3.14 or its minimal representation proposed in Figure 3.15. This new standard problem $P_{ASF,\int}$ involves new tuning parameters λ_i $i = 1, \cdots n$ (for each dof) corresponding to the pulsation below which the integral action must be effective and n new measurements corresponding to the integral of the position.

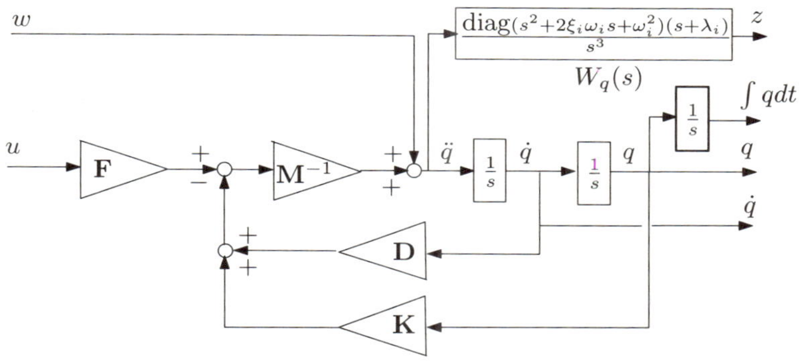

Figure 3.14. *H_∞ standard problem $P_{ASF,\int}$ based on the acceleration sensitivity function augmented with integrators*

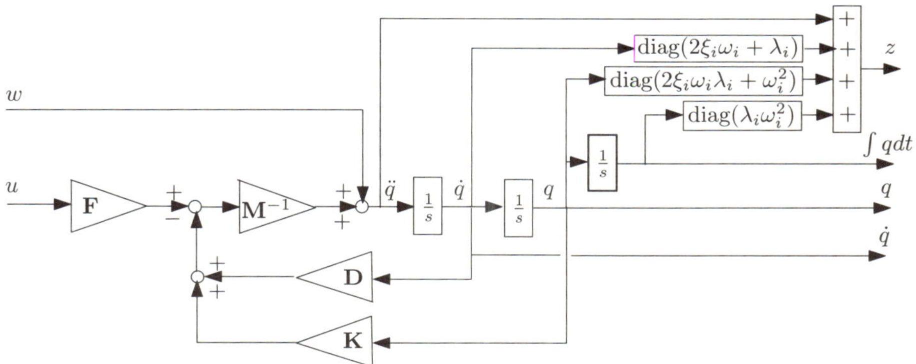

Figure 3.15. *Minimal realization of standard problem depicted in Figure 3.14*

Note that the reference input q_{ref} does not appear in this pure disturbance rejection standard problem. The reference input q_{ref} and the additional integrators are taken into account in the implementation of the optimal controller according to Figure 3.16 where $K_\infty(s)$ is the optimal $(3\,n \times m)$ H_∞ controller designed on the standard problem $P_{ASF,\int}$. In this way, the controller structure with an explicit integrator is taken into account in the standard problem. Note also that the minimal problem in Figure 3.15 is completely stabilizable and detectable.

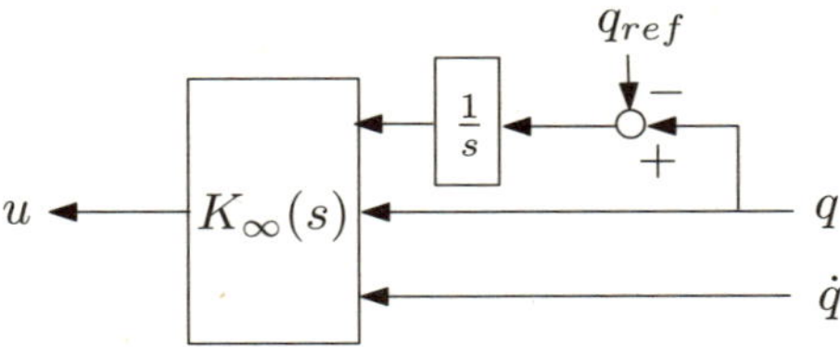

Figure 3.16. *Implementation of the controller $H_\infty(s)$ with reference input and integral term*

3.4.10. *Illustration*

The previous example $G_0 = \frac{1}{s^2-1}$ is considered again to design a output feedback (only the position q is measured) with an integral action according to the H_∞ standard problem $P_3(s)$ depicted in Figure 3.17. A first-order controller $K_{struct}(s)$ is designed using an LMI-based H_∞ solver and a reduction (from order 3 to 1). This controller is then implemented according to Figure 3.18. Step responses of the output $y(t)$ to the disturbance w and to the reference input q_{ref} are depicted in Figure 3.19 (solid lines). One can check that the disturbance rejection and the reference input tracking are perfect in steady state. The responses obtained with the initial controller $K_0(s) = -\frac{23s+32}{s+12}$ are also plotted (dashed lines) for comparison, that is step responses of $G_0/(1 - K_0G_0)$ and $-K_0G_0/(1 - K_0G_0)$. The corresponding Matlab® sequence is given below:

```
>> [a,b,c,d]=linmod('std_asf_4');
>> std=minreal(ss(a,b,c,d));
3 states removed.
>> [Kinf,CL_1,gam]=hinfsyn(std,2,1,'METHOD','lmi');
>> K_struct=red_fast(Kinf,-1e4);
>> CL_lmi=lft(std,K_struct);
>> norm(CL_lmi,'inf')      % Reduction is OK !

ans =

    1.0011
```

```
>> [aa,bb,cc,dd]=linmod('std_asf_bf');
>> CL=ss(aa,bb,cc,dd);
>> % Response to disturbance:
>> figure
>> step(CL(1,1),10)
>> hold on
>> step(feedback(G0,-K0),10);
>> % Response to reference input:
>> figure
>> step(CL(1,2),10)
>> hold on
>> step(feedback(G0*(-K0),1),10)
>> damp(CL)

        Eigenvalue                Damping       Freq. (rad/s)

 -9.88e-01                       1.00e+00         9.88e-01
 -1.04e+00 + 9.66e-01i           7.34e-01         1.42e+00
 -1.04e+00 - 9.66e-01i           7.34e-01         1.42e+00
 -2.81e+03                       1.00e+00         2.81e+03
```

Note also that the dominant closed-loop dynamics is correctly assigned.

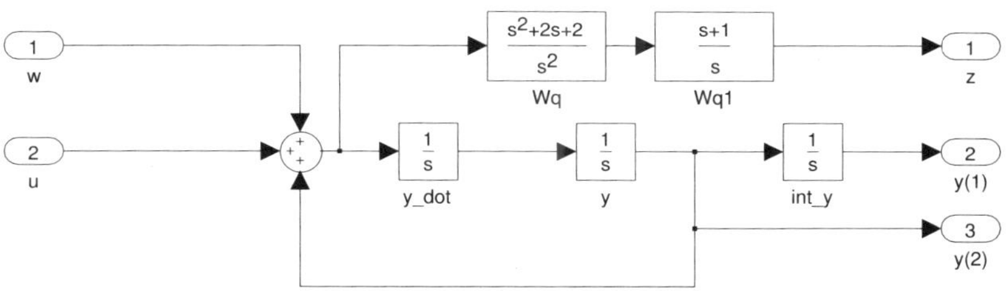

Figure 3.17. *H_∞ standard problem $P_3(s)$ (file `std_asf_4.mdl`)*

REMARK 3.1.– *To design a reduced-order controller (a first order for this application), it is recommended to design it directly using fixed-structure H_∞ synthesis rather than to design a full-order controller and then to reduce it. A possible Matlab® sequence reads (from the previous one):*

```
K_1th=ltiblock.ss('order1',1,1,2);
pb_1=lft(std,K_1th);
[CL,gam]=hinfstruct(pb_1);
K_struct=ss(CL.Blocks.order1);
```

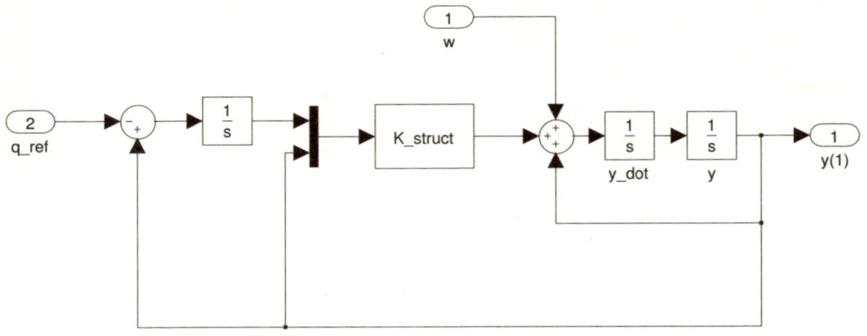

Figure 3.18. *Closed-loop system $G_{cl}(s)$ with integral term, reference input q_{ref} and disturbance w (file* `std_asf_bf.mdl`*)*

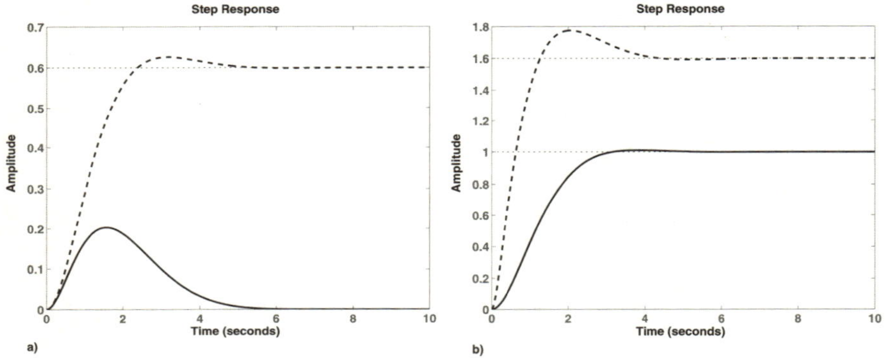

Figure 3.19. *a) Step response of $G_{cl}(s)$ from w (solid) and step response of $G_0/(1 - K_0G_0)$ (dashed). b) Step response of $G_{cl}(s)$ from q_{ref} (solid) and step responses of $-K_0G_0/(1 - K_0G_0)$ (dashed)*

3.5. Mixed H_2/H_∞ design based on the acceleration sensitivity function

In this section, the basic acceleration sensitivity problem $P(s)$ is augmented by an H_2 channel between the acceleration disturbance w and the control signal u (Figure 3.20). Such a channel is commonly introduced to minimize consumption. In other words, among all static state-feedback fitting the template on the acceleration sensitivity function, we seek the one that minimizes the energy consumption. The literal solution of this H_2/H_∞ problem was derived in the very simple case of a one dof mechanical system (see also [ALA 08]). Different cases can be considered according to the shape of the weighted open-loop acceleration sensitivity function. Such a literal solution could also be interesting to evaluate conservatism of numerical solvers proposed to solve such a H_2/H_∞ problem. Although the one dof case could seem very simple, it allows us to isolate some typical behaviors and to derive conditions for the nominal solution $\mathbf{K_0}$ to be the optimal solution of the H_2/H_∞ problem. These conditions are also recovered in the multi-variable case. The multi-variable case, when the specification is to improve the acceleration disturbance rejection ratio over the whole frequency range, is also investigated using the parametrization of all stabilizing state-feedback (see [GAD 06] and Appendix A3.3).

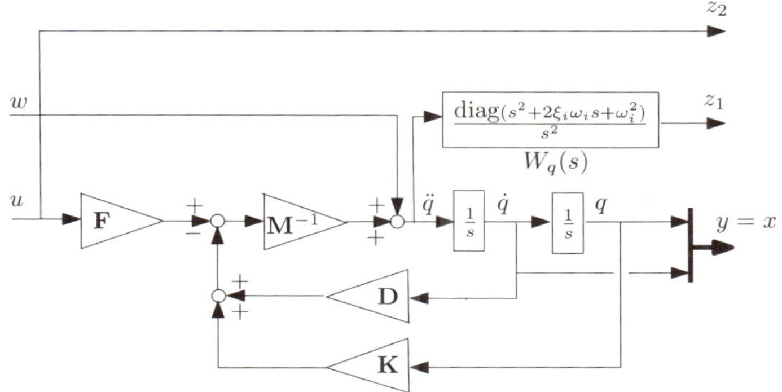

Figure 3.20. *H_2/H_∞ standard problem $P_m(s)$ for acceleration sensitivity control and energy minimization*

From equation [3.10], one can state that: $\forall\ \omega, \quad \sigma_{max}(F_l(P(j\omega), \mathbf{K_0}) = 1$, that is the closed-loop acceleration sensitivity frequency-domain response saturates the template $1/W_q(s)$ for all frequencies. Note that it is possible to find some other optimal static or dynamic stabilizing controllers $K(s)$ such that $\|F_l(P(s), K(s))\|_\infty = 1$ but also such that the template is not saturated for some frequency ω. We want to isolate the static state-feedback that minimizes the H_2 norm of the transfer between the disturbance w and $z_2 = u$.

PROBLEM 3.1.– *Considering the standard problem depicted in Figure 3.20, find a stabilizing static state-feedback controller $\widehat{K}$ such that:*

– $\|F_l(P_m(s), \widehat{K})_{w\to z_1}\|_\infty = 1$;

– $\widehat{K} = arg\min_K \|F_l(P_m(s), K)_{w\to z_2}\|_2$.

The various cases, according to the shape of the weighted open-loop acceleration sensitivity function, are reported in section 3.5.2 and illustrated on a one dof mechanical system.

3.5.1. *The one degree of freedom case*

In this section, the mixed H_2/H_∞ problem 3.1 is considered and applied to a one dof mechanical system. The optimal state-feedback gain is calculated analytically.

Only the first-order optimality conditions (necessary conditions) are derived to characterize an "admissible solution" but graphical analyses allow us to confirm that the admissible local optimum is also a global minimum. This solution is compared with numerical solution provided by mixed H_2/H_∞ synthesis in Matlab® (see section 3.5.3).

In the one dof case, all matrices **F**, **M**, **D** and **K** become scalar and will be denoted f, m, k and d, respectively. Without loss of generality we can assume that $f = 1$. We will also assume that $m > 0$ but k and d can be positive or negative (stable or unstable second-order system). The specifications ξ_1 and ω_1 are also assumed to be positive. Then, the acceleration sensitivity function has to fit the template defined by $1/W(s)$ with:

$$W(s) = \frac{s^2 + 2\xi_1\omega_1 s + \omega_1^2}{s^2}$$

and equation [3.9] becomes:

$$P(s) := \begin{bmatrix} \dot{q} \\ \ddot{q} \\ \hline z_1 \\ \hline q \\ \dot{q} \end{bmatrix} = \left[\begin{array}{cc|c|c} 0 & 1 & 0 & 0 \\ -k/m & -d/m & 1 & 1/m \\ \hline -k/m + \omega_1^2 & -d/m + 2\xi_1\omega_1 & 1 & 1/m \\ \hline 1 & 0 & 0 & 0 \\ 0 & 1 & 0 & 0 \end{array}\right] \begin{bmatrix} q \\ \dot{q} \\ \hline w \\ \hline u \end{bmatrix} \qquad [3.12]$$

Then, the open loop transfer from w to z_1 reads:

$$P_{w \to z_1}(s) = \frac{s^2 + 2\xi_1\omega_1 s + \omega_1^2}{s^2 + d/m\, s + k/m} \qquad [3.13]$$

Let us consider the static state-feedback control law $u = k_p q + k_d \dot{q} = [k_p \quad k_d]x$, then the closed-loop transfer from w to z_1 reads:

$$F_l(P(s), [k_p \quad k_d]) = F_l(P_m(s), [k_p \quad k_d])_{w \to z_1} = \frac{s^2 + 2\xi_1\omega_1 s + \omega_1^2}{s^2 + \frac{d - k_d}{m} s + \frac{k - k_p}{m}}$$

and the closed-loop transfer from w and u (or the output z_2 of the problem $P_m(s)$ depicted in Figure 3.20) reads:

$$F_l(P_m(s), [k_p \quad k_d])_{w \to z_2} = \frac{k_d s + k_p}{s^2 + \frac{d - k_d}{m} s + \frac{k - k_p}{m}}$$

Then, we can derive the objective function $J(k_p, k_v)$ and constraints as functions of the two decision variables k_p and k_v and the problem parameters m, k, d, ω_1 and ξ_1.

3.5.1.1. *Constraints*

– Stability: obviously the closed-loop system is internally stable if and only if:

$$k_p < k \qquad [3.14]$$

$$\text{and} \quad k_d < d \qquad [3.15]$$

– $\|F_l(P(s), [k_p \quad k_d])\|_\infty = \|F_l(P_m(s), [k_p \quad k_d])_{w \to z_1}\|_\infty = 1$: this constraint leads to (once the previous stability constraint is satisfied):

$$|F_l(P(j\omega), [k_p \quad k_d])| \leq 1, \ \forall\, \omega$$

$$\Longleftrightarrow \left| \frac{\omega_1^2 - \omega^2 + j2\xi_1\omega_1\omega}{\frac{k-k_p}{m} - \omega^2 + j\frac{d-k_d}{m}\omega} \right| \leq 1, \ \forall\, \omega$$

$$\Longleftrightarrow (\omega_1^2 - \omega^2)^2 + 4\xi_1^2\omega_1^2\omega^2 \leq \left(\frac{k-k_p}{m} - \omega^2\right)^2 + \left(\frac{d-k_d}{m}\right)^2 \omega^2, \ \forall\, \omega$$

$$\Longleftrightarrow \omega^2 \left[-2\omega_1^2 + 2\frac{k-k_p}{m} + 4\xi_1^2\omega_1^2 - \left(\frac{d-k_d}{m}\right)^2 \right] + \omega_1^4 - \left(\frac{k-k_p}{m}\right)^2 \leq 0, \ \forall\, \omega \qquad [3.16]$$

$$\Longleftrightarrow \frac{|k-k_p|}{m} \geq \omega_1^2, \text{ and} \qquad [3.17]$$

$$-2\omega_1^2 + 2\frac{k-k_p}{m} + 4\xi_1^2\omega_1^2 - \left(\frac{d-k_d}{m}\right)^2 \leq 0, \qquad [3.18]$$

Constraints [3.14], [3.15], [3.17] and [3.18] can be gathered into two following constraints:

$$\boxed{k_p \leq k - m\,\omega_1^2} \qquad [3.19]$$

$$\text{and} \quad \boxed{k_d \leq d - m\sqrt{4\xi_1^2\omega_1^2 - 2\omega_1^2 + 2\frac{k-k_p}{m}}} \qquad [3.20]$$

Note that this set of constraints defines a non-convex domain in the (k_p, k_v)-plane.

Considering that $P_{w\to z_1}(s) = F_l(P(s), [0 \quad 0])$, the magnitude $|P_{w\to z_1}(j\omega)|$ can be compared to 1 ($0dB$) through the sign of:

$$\varphi(\omega^2) = \omega^2 \left[-2\omega_1^2 + 2\frac{k}{m} + 4\xi_1^2\omega_1^2 - \frac{d^2}{m^2}\right] + \omega_1^4 - \frac{k^2}{m^2}, \qquad [3.21]$$

that is:

$$|P_{w\to z_1}(j\omega)| \leq 1 \iff \varphi(\omega^2) \leq 0 \qquad [3.22]$$

3.5.1.2. *Objective function*

A state space realization of $F_l(P_m(s), [k_p \quad k_d])_{w\to z_2}$ reads:

$$F_l(P_m(s), [k_p \quad k_d])_{w\to z_2} : = \left[\begin{array}{cc|c} 0 & 1 & 0 \\ (k_p - k)/m & (k_d - d)/m & 1 \\ \hline k_p & k_d & 0 \end{array}\right]$$

Therefore, using a Lyapunov equation [ZHO 96], we can calculate the objective function:

$$\boxed{J(k_p, k_v) = \|F_l(P_m(s), [k_p \quad k_d])_{w\to z_2}\|_2^2 = \tfrac{m}{2(d-k_d)}\left(k_d^2 + \tfrac{m}{k-k_p}k_p^2\right)} \qquad [3.23]$$

3.5.2. *First-order optimality conditions*

The Valentine extension of the Kuhn and Tucker approach [BOY 03] can be used to solve this optimization problem with inequality constraints. Using this approach, the Lagrange function involves squared slack variables v_1 and v_2 (associated with the two inequality constraints) [3.19] and [3.20] in addition to the two classical Lagrange multipliers λ_1 and λ_2. This Lagrange function reads in this case:

$$\begin{aligned}\mathcal{L}(k_p, k_d, \lambda_1, \lambda_2, v_1, v_2) = {} & \frac{m}{2(d-k_d)}\left(k_d^2 + \frac{m}{k-k_p}k_p^2\right) \qquad [3.24] \\ & + \lambda_1(k_p - k + m\omega_1^2 + v_1^2) \\ & + \lambda_2\left(k_d - d + m\sqrt{4\xi_1^2\omega_1^2 - 2\omega_1^2 + 2\frac{k-k_p}{m}} + v_2^2\right).\end{aligned}$$

Then, the six First-Order Conditions (FOCs) read:

$$\frac{\partial \mathcal{L}(...)}{\partial k_p} = \frac{m^2(2k-k_p)k_p}{2(d-k_d)(k-k_p)^2} + \lambda_1 - \lambda_2 \frac{1}{\sqrt{4\xi_1^2\omega_1^2 - 2\omega_1^2 + 2\frac{k-k_p}{m}}} = 0 \qquad [3.25]$$

$$\frac{\partial \mathcal{L}(...)}{\partial k_d} = \frac{m\left((2d-k_d)(k-k_p)k_d + mk_p^2\right)}{2(d-k_d)^2(k-k_p)} + \lambda_2 = 0 \qquad [3.26]$$

$$\frac{\partial \mathcal{L}(...)}{\partial \lambda_1} = k_p - k + m\omega_1^2 + v_1^2 = 0 \qquad [3.27]$$

$$\frac{\partial \mathcal{L}(...)}{\partial \lambda_2} = k_d - d + m\sqrt{4\xi_1^2\omega_1^2 - 2\omega_1^2 + 2\frac{k-k_p}{m}} + v_2^2 = 0 \qquad [3.28]$$

$$\frac{\partial \mathcal{L}(...)}{\partial v_1} = 2\lambda_1 v_1 = 0 \qquad [3.29]$$

$$\frac{\partial \mathcal{L}(...)}{\partial v_2} = 2\lambda_2 v_2 = 0 \qquad [3.30]$$

Two last FOCs allow us to manage the saturation of inequality constraints:

– if $\lambda_i = 0$, (inequality constraint i is not saturated), then v_i^2 must be positive,

– if $\lambda_i \geq 0$ (inequality constraint i is saturated), then $v_i = 0$,

and four cases must be investigated: (1) no saturated constraint, (2) one constraint is saturated ($\times$ 2) and (3) all constraints are saturated.

In each case, the conditions for the solution to be admissible as a local optimum are interpreted on the frequency response of the open-loop transfer $P_{w \to z_1}(s)$ of the standard problem P between w and z_1, that is the weighted acceleration sensibility function defined by equation [3.13].

3.5.2.1. *Case #1: no saturated constraints:* $\lambda_1 = \lambda_2 = 0$

Three sub-cases appear solving FOCs [3.25–3.28]:

– $k \geq 0$ and $d \geq 0$, then $\boxed{\widehat{k_p} = 0}$ and $\boxed{\widehat{k_d} = 0}$;

– $k \geq 0$ and $d < 0$ (unstable open-loop system), then $\boxed{\widehat{k_p} = 0}$ and $\boxed{\widehat{k_d} = 2d}$;

– $k < 0$ (unstable open-loop system), then $\boxed{\widehat{k_p} = 2\,k}$ and $\boxed{\widehat{k_d} = d - \sqrt{d^2 - 4\,m\,k}}$ (that works if $d \geq 0$ or $d < 0$).

In the three sub-cases, the conditions for $v_1^2 \geq 0$ and $v_2^2 \geq 0$ read:

$$\frac{|k|}{m} \geq \omega_1^2 \qquad [3.31]$$

$$\frac{d^2}{m^2} - 4\xi_1^2\omega_1^2 + 2\omega_1^2 - \frac{2k}{m} \geq 0 \qquad [3.32]$$

PROPOSITION 3.1.– *This solution is admissible (i.e. constraints [3.31] and [3.32] are met) if and only if:*

$$\boxed{|P_{w \to z_1}(j\omega)| \leq 1,\ \forall\,\omega} \qquad [3.33]$$

PROOF.–

$$\begin{aligned}
|P_{w \to z_1}(j\omega)| \leq 1,\ \forall\,\omega \iff\ & \varphi(\omega^2) \leq 0,\ \forall\,\omega \quad \text{(from [3.22])} \\
\iff\ & -2\omega_1^2 + 2\frac{k}{m} + 4\xi_1^2\omega_1^2 - \frac{d^2}{m^2} \leq 0 \\
& \text{and } \omega_1^4 \leq \frac{k^2}{m^2} \quad \text{(from [3.21])} \\
\iff\ & \text{[3.31] and [3.32]}
\end{aligned}$$

– *Practical interpretation:* This result is quite obvious: if the open-loop system meets the H_∞ constraint, then the best solution (from the consumption reduction point of view) is the well-known minimum energy solution, that is:

– $u = 0$ if the open-loop system is stable;

– $u \neq 0$ if the open-loop system is unstable and the optimal solution $u = [\widehat{k_p} \quad \widehat{k_d}]x$ assigns the unstable poles to stable poles symmetric w.r.t. the imaginary axis. Then, the frequency-response of the open-loop and the closed-loop, (between w and z_1) are the same and so the H_∞ constraint is met in closed-loop, although $u \neq 0$.

3.5.2.2. *Case #2: only the first constraint is saturated:* $v_1 = \lambda_2 = 0$

Then, from FOC [3.27]:

$$\boxed{\widehat{k_p} = k - m\,\omega_1^2} \qquad [3.34]$$

From FOC [3.26], we can derive:

$$\omega_1^2\,k_d^2 - 2\,d\,\omega_1^2\,k_d - (k - m\,\omega_1^2)^2 = 0 \qquad [3.35]$$

The only stabilizing solution is:

$$\boxed{\widehat{k_d} = d - \sqrt{d^2 + \left(\frac{k - m\,\omega_1^2}{\omega_1}\right)^2}} \qquad [3.36]$$

Note that this solution is independent of the parameter ξ_1.

PROPOSITION 3.2.– *Solutions [3.34] and [3.36] admissible if and only if:*

$$\boxed{|P_{w\to z_1}(j\omega_1)| < 1 \quad \textit{and} \quad |P_{w\to z_1}(0)| \geq 1} \qquad [3.37]$$

PROOF.– Reporting [3.34] in FOC [3.28]:

$$k_d - d + 2\xi_1\,\omega_1\,m + v_2^2 = 0 \qquad [3.38]$$

then, the condition $v_2^2 > 0$ reads:

$$\begin{aligned} & d^2 + \left(\frac{k - m\,\omega_1^2}{\omega_1}\right)^2 > 4\xi_1^2\,\omega_1^2\,m^2 \\ \iff\; & \frac{d^2}{m^2} - 4\xi_1^2\,\omega_1^2 + \omega_1^2 + \frac{k^2}{m^2\,\omega_1^2} - 2\frac{k}{m} > 0 \\ \iff\; & \omega_1^2\left(-2\omega_1^2 + 2\frac{k}{m} + 4\xi_1^2\,\omega_1^2 - \frac{d^2}{m^2}\right) + \omega_1^4 - \frac{k^2}{m^2} < 0 \qquad [3.39] \\ \iff\; & \text{(from [3.21]} \quad \varphi(\omega_1^2) < 0 \iff |P_{w\to z_1}(j\omega_1)| < 1 \end{aligned}$$

From FOC [3.25], the condition $\lambda_1 \geq 0$ reads:

$$\lambda_1 = -\frac{m^2(m\,\omega_1^2 + k)(k - m\,\omega_1^2)}{2\sqrt{d^2 + \left(\frac{k - m\,\omega_1^2}{\omega_1}\right)^2}(k - k_p)^2} \geq 0$$

$$\iff m^2\,\omega_1^4 - k^2 \geq 0$$

$$\iff \text{(from [3.21])} \quad \varphi(0) \geq 1 \iff |P_{w\to z_1}(0)| \geq 1$$

– *Practical interpretation:* In this case, the condition $|P_{w\to z_1}(0)| \geq 1$ means that the disturbance rejection ratio on the steady-state (DC) gain of the acceleration sensitivity function must be increased in closed-loop w.r.t. the open-loop (the disturbance rejection performance must be increased in low frequency). While the condition $|P_{w\to z_1}(j\omega_1)| < 1$ means that for pulsations approximately ω_1, the open-loop disturbance rejection performance is better than the specification $1/W_q$ introduced in the standard problem $P(s)$ (see Figure 3.2). That is illustrated in the following numerical example.

– *Numerical illustration:* Let us consider the following numerical application:

$$m = 10\,\text{kg}, \quad d = 100\,\text{Ns/m}, \quad k = 10\,\text{N/m}, \quad \omega_1 = 2\,\text{rad/s}, \quad \xi_1 = 0.5 \qquad [3.40]$$

The frequency-domain response (magnitude plot) of $P_{w\to z_1}(s)$ is depicted in Figure 3.21 (solid grey line). Thus, conditions [3.37] are met and the admissible solution (equations [3.34] and [3.36]) reads:

$$\widehat{k_p} = -30\,\text{N/m}, \quad \widehat{k_d} = -1.1187\,\text{Ns/m} \quad \text{and} \quad J(\widehat{k_p}, \widehat{k_d}) = 11.2$$

The closed-loop frequency-domain response $|F_l(P(j\omega), [\widehat{k_p} \quad \widehat{k_d}])|$ is also depicted in Figure 3.21 (solid black line): we can notice that the specification is met:

$$|F_l(P(j\omega), [\widehat{k_p} \quad \widehat{k_v}])| \leq 1 \quad \forall\,\omega$$

But the optimal control (from the energy minimum point of view) acts only in low frequency and recovers the open-loop behavior in the frequency range (around ω_1 and toward infinity) where the open-loop rejection performance is good enough (and better than the rejection performance prescribed by $1/W_q(s) = \frac{s^2}{s^2+2\xi_1\,\omega_1\,s+\omega_1^2}$).

The Second-Order Condition (SOC) (i.e. the Hessian $\nabla^2 J(\widehat{k_p}, \widehat{k_d})$ is positive) is not detailed. This SOC is required to conclude that this admissible solution is a local minimum (due to the non-convexity of constraints and the objective function) but the 3D plot presented in Figure 3.22, allows us to conclude that (for this numerical application) this admissible solution is also a global minimum.

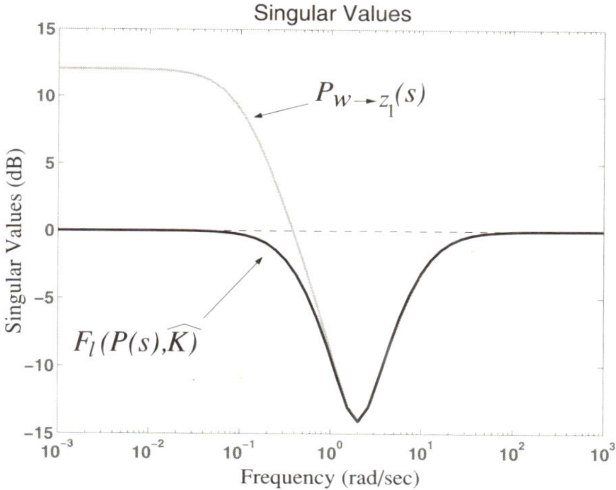

Figure 3.21. *Open-loop* $|P_{w\to z_1}(j\omega)|$ *and closed-loop* $|F_l(P(\omega), [\widehat{k_p} \quad \widehat{k_d}])|$ *frequency-domain responses of the* H_∞ *performance channel (case #2)*

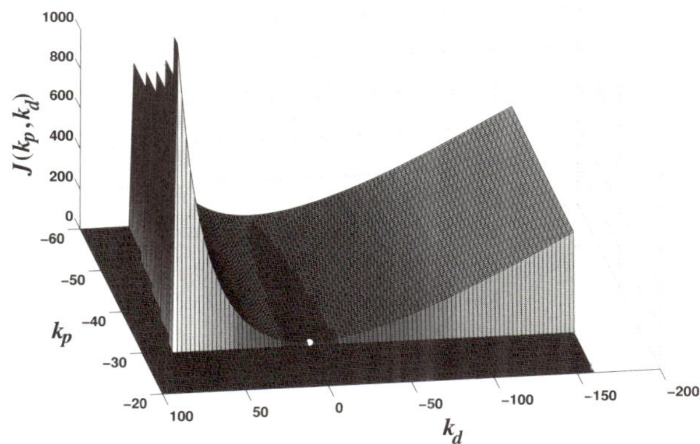

Figure 3.22. *Three-dimensional visualization on the objective function* $J(k_p, k_d)$ *and the constraints [3.19] and [3.20] in the plane* (k_p, k_d)*, the white* $*$ *corresponds to the optimal solution (case #2):* $(\widehat{k_p}, \widehat{k_d}, J(\widehat{k_p}, \widehat{k_d}))$

3.5.2.3. *Case #3: only the second constraint is saturated:* $v_2 = \lambda_1 = 0$

From FOC [3.28], $\widehat{k_d}$ can be expressed as a function of k_p:

$$\boxed{\widehat{k_d} = d - m\sqrt{4\xi_1^2\,\omega_1^2 - 2\omega_1^2 + 2\left(\frac{k-k_p}{m}\right)}}. \qquad [3.41]$$

Then, reporting this equation in FOC [3.25] allows us to express λ_2 as a function of k_p:

$$\lambda_2 = \frac{m(2k - k_p)k_p}{2(k - k_p)^2}. \qquad [3.42]$$

Substituting k_d and λ_2 by equations [3.41] and [3.42] in FOC [3.26] allows us to find an equation in k_p. After some (tedious) developments, $\widehat{k_p}$ is solution of the third-order polynomial equation:

$$\begin{aligned}\mathcal{P}_3(k_p) = 3m\, k_p^3 &+ (4m^2\, \omega_1^2 + d^2 - 11m\, k - 8m^2\, \xi_1^2\, \omega_1^2)\, k_p^2 \\ &+ (10m\, k^2 + 16m^2\, k\, \xi_1^2\, \omega_1^2 - 8m^2\, k\, \omega_1^2 - 2k\, d^2)\, k_p \\ &+ (k^2\, d^2 + 2m^2\, k^2\, \omega_1^2 - 4m^2\, k^2\, \xi_1^2\, \omega_1^2 - 2m\, k^3) = 0.\end{aligned} \qquad [3.43]$$

The condition $\lambda_2 \geq 0$ reads (from [3.42]): $(2k - k_p)k_p \geq 0$.

The condition $v_1^2 > 0$ reads (from [3.27]): $k_p < k - m\,\omega_1^2$.

That is: $|k|/m > \omega_1^2$ or $|P_{w\to z_1}(0)| < 1$ and

– $0 \leq k_p < k - m\,\omega_1^2$ if $k \geq 0$;

– $2\,k \leq k_p < k - m\,\omega_1^2$ if $k < 0$.

Then, a sufficient condition for [3.43] to have a solution in $k_p \in [0, \quad k - m\,\omega_1^2[$ (respectively $[2k, \quad k - m\,\omega_1^2[$) for $k > 0$ (respectively $k < 0$) is:

$$|P_{w\to z_1}(j\omega)| \geq 1, \text{ for } \omega \to \infty \quad \text{and} \quad |P_{w\to z_1}(j\frac{\omega_1}{\sqrt{1 + 4\xi_1^2}})| < 1 \qquad [3.44]$$

PROOF.– ***(in the case*** $k > 0$***)***: Considering that $0 \leq k_p \leq k - m\,\omega_1^2$, we can evaluate the polynomial $\mathcal{P}_3(k_p)$ for extremal values 0 and $k - m\,\omega_1^2$:

$$\begin{aligned}\mathcal{P}_3(0) &= k^2 m^2 (2\omega_1^2 - 2\frac{k}{m} - 4\xi_1\omega_1^2 + \frac{d^2}{m^2}) \\ &= -k^2 m^2 \varphi(\omega^2)|_{\omega\to\infty} \quad \text{(from: [3.21])}\end{aligned}$$

$$\begin{aligned}\mathcal{P}_3(k - m\,\omega_1^2) &= m^2k^2\omega_1^2 + 4\xi_1^2 m^2k^2\omega_1^2 - 2m^3k\omega_1^4 + m^4\omega_1^6 - 8\xi_1^2 m^4\omega_1^6 + d^2m^2\omega_1^4 \\ &= m^4\omega_1^4\left(\frac{k^2}{\omega_1^2 m^2}(1 + 4\xi_1^2) - 2\frac{k}{m} + \omega_1^2 - 8\xi_1^2\omega_1^2 + \frac{d^2}{m^2}\right) \\ &= m^4\omega_1^4\left(2\omega_1^2 - 2\frac{k}{m} - 4\xi_1\omega_1^2 + \frac{d^2}{m^2} + \frac{1 + 4\xi_1^2}{\omega_1^2}(\frac{k^2}{m^2} - \omega_1^4)\right) \\ &= -m^4\omega_1^2(1 + 4\xi_1^2)\varphi\left(\frac{\omega_1^2}{1 + 4\xi_1^2}\right) \quad \text{(from: [3.21])}\end{aligned}$$

If $\varphi(\omega^2)|_{\omega\to\infty}$ and $\varphi\left(\frac{\omega_1^2}{1+4\xi_1^2}\right)$ have opposite sign, then $\mathcal{P}_3(0)$ and $\mathcal{P}_3(k - m\,\omega_1^2)$ have opposite sign and $\mathcal{P}_3(k_p) = 0$ has a solution in $[0,\ k - m\,\omega_1^2[$. Since $\varphi(0) < 0$ ($|P_{w\to z_1}(0)| < 1$), a sufficient condition reads:

$$\varphi\left(\frac{\omega_1^2}{1+4\xi_1^2}\right) < 0 \iff |P_{w\to z_1}(j\frac{\omega_1}{\sqrt{1+4\xi_1^2}})| < 1 \text{ and}$$
$$\varphi(\omega^2)|_{\omega\to\infty} \geq 0 \iff |P_{w\to z_1}(j\omega)| \geq 1, \text{ for } \omega \to \infty$$

Considering:

– the general form of $P_{w\to z_1}(s)$ given in [3.13],

– $\lim_{\omega\to\infty} |P_{w\to z_1}(j\omega)| = 1$,

– $|k|/m \geq \omega_1^2$ (in this case),

we can state the following proposition for this special case:

PROPOSITION 3.3.– *The solution* $\widehat{k_p}$ *of equation [3.43] in* $[0,\ \ k - m\,\omega_1^2]$ *(respectively* $[2\,k,\ \ k - m\,\omega_1^2]$*) if* $k \geq 0$ *(respectively* $k < 0$*) and the corresponding* $\widehat{k_d}$ *(given by equation [3.41]) is an admissible solution if:*

$$\boxed{|P_{w\to z_1}(j\omega)| \geq 1,\ \forall\omega \geq \sqrt{\tfrac{k}{m}} \quad \text{and} \quad |P_{w\to z_1}(j\omega)| \leq 1,\ \forall\omega \leq \tfrac{\omega_1}{\sqrt{1+4\xi_1^2}}}. \qquad [3.45]$$

– *Practical interpretation:* In this case, the first part of the condition [3.45] means that the disturbance rejection ratio must be increased in high-frequency and the second part means that the low-frequency response of the open-loop acceleration sensitivity function meets the specification $1/W_q$. This is illustrated in the following numerical example.

– *Numerical illustration:* Let us consider the following numerical application (unstable open-loop system):

$$m = 10\,\text{kg}, \quad d = -1\,\text{Ns/m}, \quad k = 10\,\text{N/m}, \quad \omega_1 = 0.2\,\text{rad/s}, \quad \xi_1 = 0.7 \qquad [3.46]$$

The frequency-domain response (magnitude plot) of $P_{w\to z_1}(s)$ is depicted in Figure 3.23 (solid grey line). Thus, conditions [3.45] are met and the optimal solution reads:

$$\widehat{k_p} = 2.781\,\text{N/m}, \quad \widehat{k_d} = -13.01\,\text{Ns/m} \quad \text{and} \quad J(\widehat{k_p}, \widehat{k_d}) = 74.9$$

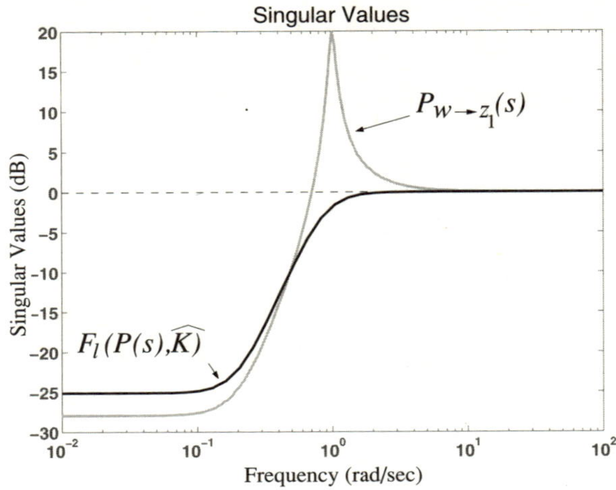

Figure 3.23. *Open-loop $|P_{w\to z_1}(j\omega)|$ and closed-loop $|F_l(P(j\omega),[\widehat{k_p} \quad \widehat{k_d}])|$ frequency-domain responses of the H_∞ performance channel (case #3)*

The closed-loop frequency-domain response $|F_l(P(j\omega),[\widehat{k_p} \quad \widehat{k_d}])|$ is also depicted in Figure 3.23 (solid black line): we can notice that the specification is met:

$$|F_l(P(j\omega),[\widehat{k_p} \quad \widehat{k_v}])| \leq 1 \quad \forall \omega$$

But the optimal (from the energy minimum point of view) control acts mainly in high frequency and (what is not intuitive) degrades lightly the open-loop rejection performance in the low-frequency response.

As in the previous case, Figures 3.24 and 3.25 (zoom) allow us to conclude that (for this numerical application) this admissible solution is also a global minimum.

3.5.2.4. *Case #4: both constraints* [3.19] *and* [3.20] *are saturated:* $v_1 = v_2 = 0$

From FOCs [3.27] and [3.28],

$$\boxed{\widehat{k_p} = k - m\,\omega_1^2}, \qquad [3.47]$$

$$\boxed{\widehat{k_d} = d - 2m\xi_1\omega_1}, \qquad [3.48]$$

and we can recognize the nominal solution $\mathbf{K_0}$ (equation [3.6]) for the one dof case (with $\mathbf{F} = 1$).

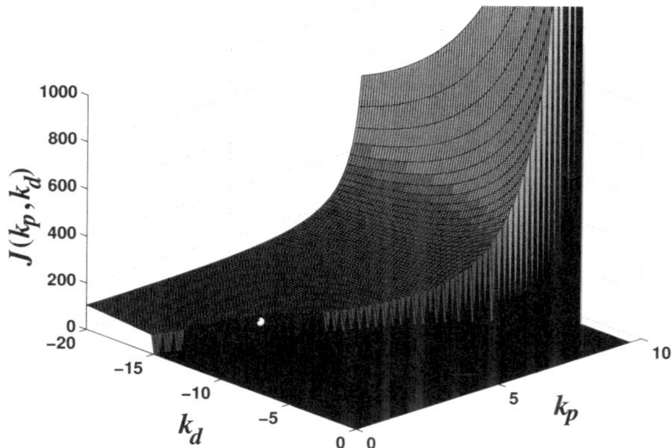

Figure 3.24. *Three-dimensional visualization on the objective function $J(k_p, k_d)$ and the constraints [3.19] and [3.20] in the plane (k_p, k_d), the white $*$ corresponds to the optimal point (case #3): $(\widehat{k_p}, \widehat{k_d}, J(\widehat{k_p}, \widehat{k_d}))$*

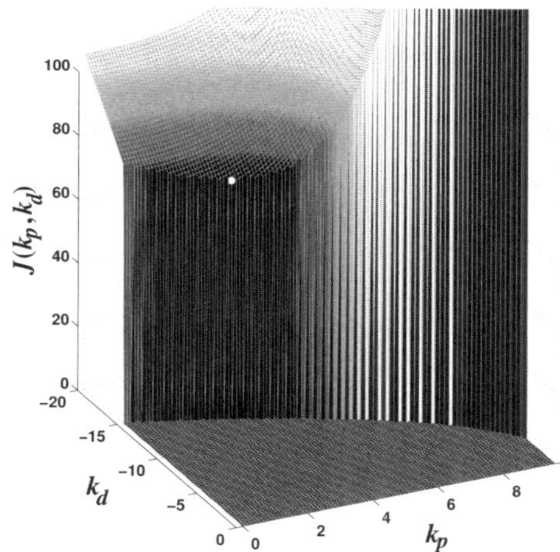

Figure 3.25. *Three-dimensional visualization on the objective function $J(k_p, k_d)$ and the constraints* [3.19] *and* [3.20] *in the plane (k_p, k_d), the white $*$ corresponds to the optimal point (case #3, zoom): $(\widehat{k_p}, \widehat{k_d}, J(\widehat{k_p}, \widehat{k_d}))$*

Then, FOC [3.26] and [3.25] leads to:

$$\lambda_2 = \frac{(4m^2\xi_1^2\omega_1^2 - d^2)\omega_1^2 - (k - m\omega_1^2)^2}{8m\xi_1^2\omega_1^4}$$

and

$$\lambda_1 = \frac{\lambda_2}{2\xi_1\omega_1} + m\frac{\omega_1^4 - \frac{k^2}{m^2}}{4\xi_1\omega_1^5}$$

λ_2 can be also expressed under the form:

$$\lambda_2 = m\frac{\omega_1^2(-2\omega_1^2 + 2\frac{k}{m} + 4\xi_1^2\omega_1^2 - \frac{d^2}{m^2}) + \omega_1^4 - \frac{k^2}{m^2}}{8\xi_1^2\omega_1^4}$$

or using [3.21]:

$$\lambda_2 = \frac{m}{8\xi_1^2\omega_1^4}\varphi(\omega_1^2)$$

Then, λ_1 can be also expressed under the form:

$$\lambda_1 = \frac{m}{16\xi_1^3\omega_1^5}\varphi(\omega_1^2) + \frac{m}{4\xi_1\omega_1^5}\varphi(0) = \frac{m(1+4\xi_1^2)}{16\xi_1^3\omega_1^5}\varphi\left(\frac{\omega_1^2}{1+4\xi_1^2}\right)$$

Then, using [3.22], $\lambda_1 \geq 0$ and $\lambda_2 \geq 0$ if and only if:

$$|P_{w\to z_1}(j\omega_1)| \geq 1 \text{ and } |P_{w\to z_1}(j\frac{\omega_1}{\sqrt{1+4\xi_1^2}})| \geq 1$$

From a practical point of view, the sufficient condition stated in the following proposition is worth mentioning:

PROPOSITION 3.4.– *The solution* $\mathbf{K_0} = [\widehat{k_p} \quad \widehat{k_d}]$ *(given by equations [3.47] and [3.48]) is an admissible solution if:*

$$\boxed{|P_{w\to z_1}(j\omega)| \geq 1 \quad \forall\omega}. \qquad [3.49]$$

Note that this condition is not a necessary condition. Indeed, the following numerical values $m = 1$ kg, $k = 1$ N/m, $d = 0$, $\xi_1 = 0.5$ and $\omega_1 = 7$ rad/s (or $m = 10$ kg, $k = 10$ N/m, $d = 1$ Ns/m, $\xi_1 = 0.7$ and $\omega_1 = 0.9$ rad/s) do not meet this condition while the solution is admissible. It can be shown that the magnitude of $\frac{s^2+7s+49}{s^2+1}|_{s=j\omega}$ is a few lower than 1 ($0\,dB$) for $\omega > 7.44$ rad/s. But such a case is quite marginal and without practical interest.

– *Practical interpretation:* This case is certainly the most interesting since, in practical problems, the performance specification consists of improving disturbance rejection ratio for all frequencies, that is the frequency response of the open-loop acceleration sensitivity function is greater that the specification $|1/W_q(j\omega)|$ for all frequencies ω. Then, the nominal solution $\mathbf{K_0}$ is the optimal solution (from the minimum energy point of view) and just saturates the H_∞ constraint, that is $|F_l(P(j\omega), \mathbf{K_0})| = 1,\ \forall\, \omega$. This is illustrated in the following numerical example.

– *Numerical illustration:* Let us consider the following numerical application:

$$m = 10\,\text{kg}, \quad d = 1\,\text{Ns/m}, \quad k = 10\,\text{N/m}, \quad \omega_1 = 2\,\text{rad/s}, \quad \xi_1 = 0.7 \quad [3.50]$$

The frequency-domain response (magnitude plot) of $P_{w\to z_1}(s)$ is depicted in Figure 3.26 (solid grey line). Thus, conditions [3.49] are met and the optimal solution reads:

$$\widehat{k_p} = -30\,\text{N/m}, \quad \widehat{k_d} = -27\,\text{Ns/m} \quad \text{and} \quad J(\widehat{k_p}, \widehat{k_d}) = 170.4$$

The closed-loop frequency-domain response $|F_l(P(j\omega), [\widehat{k_p} \quad \widehat{k_d}])|$ is also depicted in Figure 3.26 (solid black line): we can note that the specification is met:

$$|F_l(P(j\omega), [\widehat{k_p} \quad \widehat{k_v}])| \leq 1 \quad \forall\, \omega$$

and that the H_∞ constraint is saturated for all frequencies.

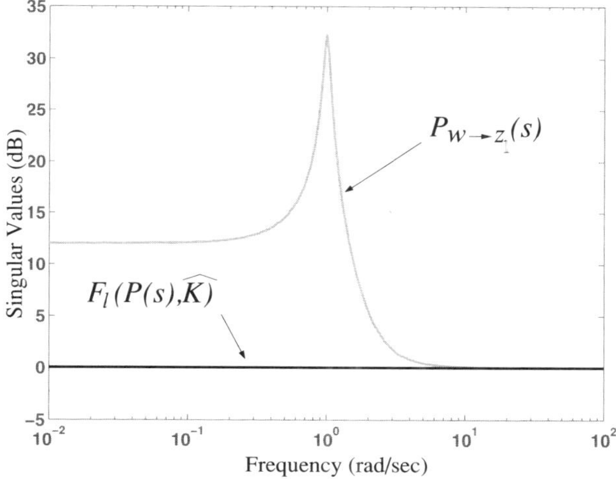

Figure 3.26. *Open-loop $|P_{w\to z_1}(j\omega)|$ and closed-loop $|F_l(P(j\omega), [\widehat{k_p} \quad \widehat{k_d}])|$ frequency-domain responses of the H_∞ performance channel (case # 4)*

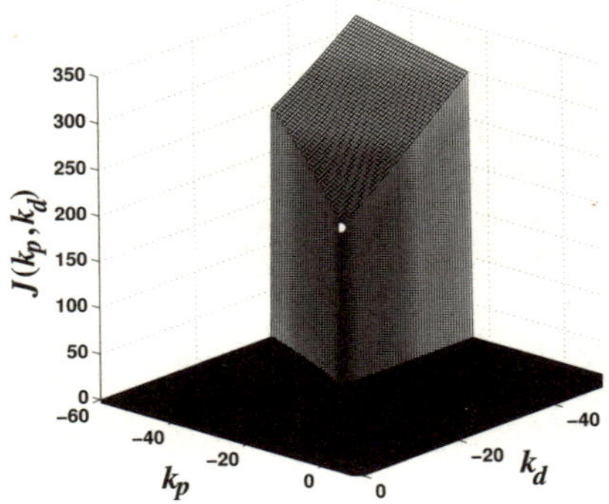

Figure 3.27. *3D visualization on the objective function $J(k_p, k_d)$ and the constraints [3.19] and [3.20] in the plane (k_p, k_d), the white $*$ corresponds to the optimal point (case # 4): $(\widehat{k_p}, \widehat{k_d}, J(\widehat{k_p}, \widehat{k_d}))$*

As in the previous case, Figures 3.27 allows us to conclude that, for this numerical application, this solution is also a global minimum.

A general function `h2hinfsol.m` that aggregates these analytical results is available at http://personnel.isae.fr/daniel-alazard/matlab-packages. The usage of this function is given in Appendix A4.9.

3.5.3. *Numerical solution using Matlab®*

The previous analytical development could seem a little bit tedious to solve a pair of proportional derivative gains for a simple second-order system. It is interesting to remark that such a solution cannot be easily recovered using numerical solvers.

REMARK 3.2.– Full-order solvers produce a dynamic controller with order equal to the plant order (2). It can be shown that the controller dynamics is very fast and so the controller can be reduced to its DC gain without any performance degradation (on both H_2 and H_∞ performance indexes).

Indeed, the MATLAB sequence required to solve this H_2/H_∞ problem using embedded macro-function `hinfmix` [GAH 94] is (once `m, d, k, xi, w` are defined):

```
P=ss([0 1;-k/m -d/m],[0 0;1 1/m],...
     [-k/m+w*w -d/m+2*xi*w;0 0;1 0;0 1],[1 1/m;0 1;0 0;0 0]);
[gopt,h2i2,kmix]=hinfmix(pck(P.a,P.b,P.c,P.d),[1 2 1],[1.001 100 0 1]);
[ak,bk,ck,dk]=unpck(kmix);kmix=ss(ak,bk,ck,dk);   % system matrix to ss
Kopt=dcgain(kmix);        % The controller is reduced to its DC gain.
CL1=lft(P,Kopt);          % Compute the closed-loop system
norm(CL1(2,1))^2          % Value of the objective function
norm(CL1(1,1),'inf')      % Check the H_\infty constraint
```

	Analytical	`hinfmix`	`hifoo`
Case # 2 (equation [3.40])	11.2	13.3	14.4
Case # 3 (equation [3.46])	74.9	98.0	152
Case # 4 (equation [3.50])	170.35	319	234

Table 3.2. *Comparison of objective function values $J(k_p, k_d)$ obtained with analytical and numerical solutions*

The solution obtained by such a numerical approach (based on LMI) on various cases studied in section 3.5.2 is shown in Table 3.2 and compared (from the objective function point of view) with the exact and literal solution. The table also presents the results that are obtained using `hifoo`. `Hifoo` [ARZ 11, GUM 09] allows a static controller to be designed directly (without *a posteriori* reduction) on such an H_2/H_∞ problem according to the following sequence (once `hifoo` is installed in the user working directory):

```
P1=P([1 3 4],[1 2]);    % Hinfty 3x2 standard problem
P2=P([2 3 4],[1 2]);    % H2 3x2 standard problem
input.U1=1;input.U2=2;output.Y1=1;output.Y2=[2 3];
set(P1,'InputGroup',input,'OutputGroup',output);
set(P2,'InputGroup',input,'OutputGroup',output);
K=hifoo({P1,P2},0,ss([0 0]),'ht',[1.01,1000]);
   % The H_infty constraint is relaxed to 1.01 (instead of 1),
   % The H2 index to be optimized is constrained to be < 1000.
CL2=lft(P,K.d);         % Compute the closed-loop system
norm(CL2(2,1))^2        % Value of the objective function
norm(CL3(1,1),'inf')    % Check the H_\infty constraint.
```

REMARK 3.3.– In order to have a stable result, an initial guess ([0 0]) is provided to `hifoo`; otherwise, the solution depends on the current run and is not reproducible.

We can see that the solution provided by numerical solvers is always suboptimal, mainly in case #4 which is more important from a practical point of view (full frequency-range disturbance rejection): this very simple H_2/H_∞ problem is still *an open benchmark* for numerical solvers.

Note that in case #4, the optimal solution can be recovered using the full-order LMI-based H_∞ solver (forgetting the H_2 channel) and a reduction of the controller to its DC gain, exactly as was done in section 3.4.2. This is illustrated by the following MATLAB sequence:

```
>> [kinf,CL4,gam]=hinfsyn(P([1 3 4],[1 2]),2,1,'METHOD','lmi');
>> kinf=dcgain(kinf)   % The controller is reduced to its DC gain.

kinf =

  -30.0001   -27.0010

>> CL3=lft(P,kinf);    % Compute the closed-loop system
>> norm(CL3(2,1))^2    % Value of the objective function

ans =

  170.3608

>> norm(CL3(1,1),'inf') % Check the H_\infty constraint

ans =

     1
```

This solution is very close to the optimum. Such a behavior works also in the multi-variable case (see illustration in section 3.6) and can be explained using the parameterization of all stabilizing state-feedbacks and LMI solvers. This is detailed in Appendix A3.3. Note that the nominal state-feedback gain $\mathbf{K_0}$ defined in equation [3.6] is optimal (from the H_2 performance index point of view) only in case #4. That is why in this case (full frequency-range disturbance rejection), although the H_2 channel is not taken into account, `hinflmi` recovers the optimal solution $\mathbf{K_0}$. More sophisticated solvers like `hinfmix` explore the set of parameterized controllers around the central controller but miss the optimal solution.

The reader can download from http://personnel.isae.fr/daniel-alazard /matlab-packages a Matlab® script file `demo_section_3_5_3.m` to run these various numerical analyses.

3.5.4. *Multi-variable case*

The literal development in the previous section was possible because, in the one dof case, the H_∞ constraint on a single input single output (SISO) second-order system can be analytically expressed. For multi-variable systems, this is no longer the case, but propositions 3.1–3.4 on the frequency response of the weighted open-loop acceleration sensitivity function can be applied using the notion of minimal and maximal singular value (σ_{min} and σ_{max}). For instance, case #4, where the rejection performance must be increased for all frequencies, will be characterized by the following proposition in the multi-variable case.

PROPOSITION 3.5.– *The solution* $\mathbf{K_0}$ *(equation [3.6]) is an admissible solution for problem 3.1 if:*

$$\boxed{\sigma_{min}(P_{w\rightarrow z_1}(j\omega)) \geq 1 \quad \forall \omega} \qquad [3.51]$$

This proposition will not be proved in this book. It is more important to know how such a solution $\mathbf{K_0}$ can be recovered using numerical tools. Indeed, when the standard problem $P(s)$ is augmented by additional specifications, it is important for the current design to be attracted by a unique solution ($\mathbf{K_0}$) and not by a set of solutions that can be very different and could introduce some discontinuities in the trade-off tuning or several local minima.

The illustration proposed in the next section shows that the main conclusions obtained in this chapter in the one dof case are still valid in the multi-variable case:

– Mixed H_2/H_∞ solvers failed to find the optimal solution and so the problem is still open.

– In case #4, full-order LMI-based H_∞ solver can be used to find the optimal state-feedback.

– The standard problem based on the acceleration sensitivity function can be augmented with actuator dynamics (for instance) and additional H_∞ specifications (roll-off) to design a reduced-order dynamic controller. Then the low-frequency behavior allows the performance specified through parameters ξ_i and ω_i to be recovered.

3.6. Aircraft lateral flight control design

3.6.1. *Model and specifications*

Let us consider the lateral flight model of a civil aircraft:

$$\begin{bmatrix} \dot{\beta} \\ \dot{p} \\ \dot{r} \\ \dot{\phi} \\ \hline n_y \\ p \\ r \\ \phi \end{bmatrix} = \left[\begin{array}{cccc|cc} -0.140 & 0.053 & -0.999 & 0.047 & 0 & 0 \\ -2.461 & -0.992 & 0.262 & 0 & 0.404 & 0.260 \\ 1.595 & -0.041 & -0.267 & 0 & 0. & -0.680 \\ 0 & 1.000 & 0.053 & 0 & 0 & 0 \\ \hline 0.0433 & -0.0003 & 0.0016 & 0 & 0.0001 & -0.0075 \\ 0 & 1 & 0 & 0 & 0 & 0 \\ 0 & 0 & 1 & 0 & 0 & 0 \\ 0 & 0 & 0 & 1 & 0 & 0 \end{array}\right] \begin{bmatrix} \beta \\ p \\ r \\ \phi \\ \hline dp \\ dr \end{bmatrix}, \qquad [3.52]$$

where:

– β is the side slip (yaw) angle, p is the roll rate, r is the yaw rate and ϕ is the bank (roll) angle ($x = [\beta \quad p \quad r \quad \phi]^T$);

– dp is the aileron deflection and dr is the rudder deflection ($u = [dp \quad dr]^T$);

– n_y is the lateral load factor (acceleration) ($y = [n_y \quad p \quad r \quad \phi]^T$).

The specifications are:

– yaw/roll decoupling;

– second-order behavior on yaw axis defined by $\omega_y = 3$ rad/s and $\xi_y = 0.7$;

– second-order behavior on roll axis defined by $\omega_r = 3$ rad/s and $\xi_r = 0.7$.

First, it can be noted that this model is not described under the generalized second-order form (equation [3.1]). To address the yaw/roll decoupling, the vector q of the two dof of interest must be defined by $q = [\beta \quad \phi]^T$. So, it is required to perform a transformation in order to define a new state variable $\widetilde{x} = [q^T \quad \dot{q}^T]^T = [\beta \quad \phi \quad \dot{\beta} \quad \dot{\phi}]^T$, that is:

$$\widetilde{x} = \begin{bmatrix} 1 & 0 & 0 & 0 \\ 0 & 0 & 0 & 1 \\ -0.140 & 0.053 & -0.999 & 0.047 \\ 0 & 1.000 & 0.053 & 0 \end{bmatrix} x$$

Note that such a transformation is not always possible. Indeed, if there is a direct term between u and $\dot{q}$ in the original representation, then it is not possible to represent the system with a second-order form in q (equation [3.1]).

Let us denote:

$$\begin{cases} \dot{\widetilde{x}} = \widetilde{A}\widetilde{x} + \widetilde{B}u \\ y = \widetilde{C}\widetilde{x} + \widetilde{D}u, \end{cases}$$

this state space realization. Then, we can choose $\mathbf{M} = I_2$ and represent the model by the second-order form:

$$\ddot{q} + \mathbf{D}\dot{q} + \mathbf{K}q = \mathbf{F}u$$

where $\mathbf{K} = -\widetilde{A}(3{:}4, 1{:}2)$, $\mathbf{D} = -\widetilde{A}(3{:}4, 3{:}4)$ and $\mathbf{F} = \widetilde{B}(3{:}4, 1{:}2)$ (where $A(i{:}j, k{:}l)$ is the submatrix of A consisted of rows i to j and columns k–l).

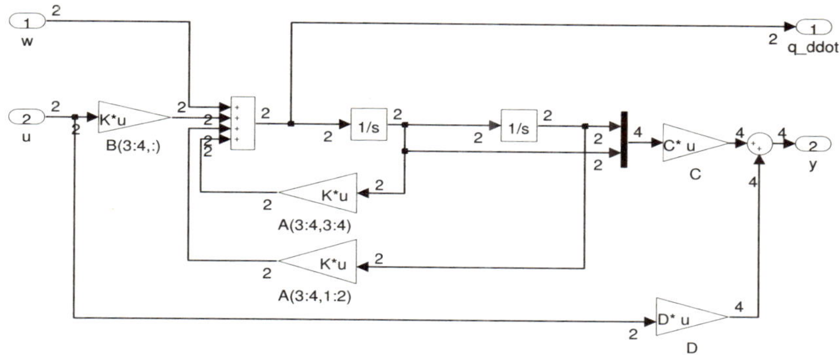

Figure 3.28. *The acceleration sensitivity problem $P_{A/C}(s)$ (without weigh $W_q(s)$)*

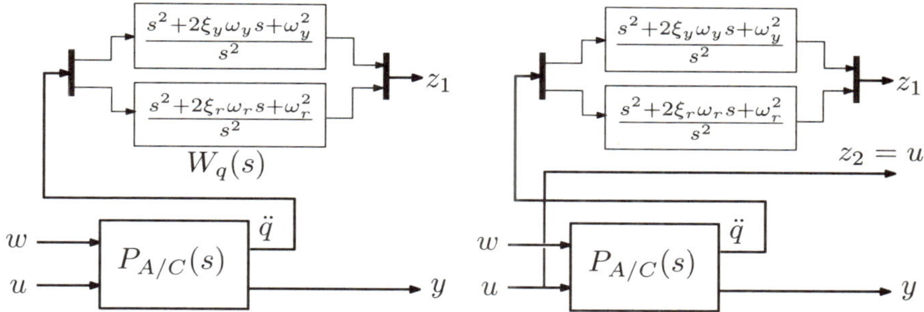

Figure 3.29. *The H_∞ standard problem $P(s)$ on the acceleration sensitivity function weighted by $W_q(s)$ (left) and the H_2/H_∞ standard problem $P_m(s)$ (right)*

3.6.2. *Basic H_2/H_∞ control problem*

Then, the acceleration sensitivity problem $P_{A/C}(s)$ with output feedback is depicted in Figure 3.28. Figure 3.29 details:

– The H_∞ problem $P(s)$ (on the left) based on $P_{A/C}(s)$ and the weight $W_q(s)$. $W_q(s)$ is directly expressed from the yaw and roll dynamics specifications.

– The H_2/H_∞ problem $P_m(s)$ (on the right) including an additional H_2 channel between w and $z_2 = u$. That is:

PROBLEM 3.2.– *Considering the standard problems depicted in Figure 3.29, find a stabilizing static output-feedback $\widehat{K_y}$ such that:*

– $\|F_l(P_m(s), \widehat{K_y})_{w\to z_1}\|_\infty = \|F_l(P(s), \widehat{K_y})\|_\infty = 1$;

– $\widehat{K_y} = \mathit{arg}\min_{K_y} \|F_l(P_m(s), K_y)_{w\to z_2}\|_2$.

The minimal state-space realization of the problem $P_m(s)$ reads:

$$P_m(s) := \left[\begin{array}{cc|c|c} 0_2 & I_2 & 0_2 & 0_2 \\ -\mathbf{K} & -\mathbf{D} & I_2 & \mathbf{F} \\ \hline \mathrm{diag}([\omega_y^2 \;\; \omega_r^2]) - \mathbf{K} & \mathrm{diag}([2\xi_y\omega_y \;\; 2\xi_r\omega_r]) - \mathbf{D} & I_2 & \mathbf{F} \\ \multicolumn{2}{c|}{0_{2\times 4}} & 0_2 & I_2 \\ \hline \multicolumn{2}{c|}{\widetilde{C}} & 0_2 & \widetilde{D} \end{array}\right].$$

The nominal state-feedback gain [3.6] is:

$$\mathbf{K_0} = \mathbf{F}^{-1}[\mathbf{K} - \mathrm{diag}([\omega_y^2 \;\; \omega_r^2]) \quad \mathbf{D} - \mathrm{diag}([2\xi_y\omega_y \;\; 2\xi_r\omega_r])]$$

$$\mathbf{K_0} = \begin{bmatrix} 11.9797 & -22.6902 & 3.8309 & -8.0726 \\ -10.8114 & 0.6819 & -5.5711 & 0.1769 \end{bmatrix}$$

and the corresponding output feedback reads:

$$K_{y0} = \mathbf{K_0}(\widetilde{C} + \widetilde{D}\mathbf{K_0})^{-1} = \begin{bmatrix} 95.6160 & -7.8506 & -0.3694 & -21.9936 \\ -83.8187 & -0.1351 & 2.1687 & -0.0327 \end{bmatrix}.$$

The frequency-domain response (singular values) of the weighted open-loop acceleration sensitivity function is depicted in Figure 3.30 (solid grey line) and highlights that the condition of proposition 3.5 is met. The acceleration disturbance rejection must be increased on the whole frequency range. Therefore, we can guess that the static output feedback K_{y0} is the optimal solution, that is $K_{y0} = \min_{K_y} \|F_l(P_m(s), K_y)_{w\to z_2}\|_2$ and K_{y0} saturates the H_∞ constraint $\|F_l(P(s), K_y)\|_\infty \leq 1$. Indeed, the frequency response of $F_l(P(s), K_{y0})$ is also depicted in Figure 3.30 (solid black line) and highlights that $\sigma_{max}(F_l(P(j\omega), K_{y0})) = \sigma_{min}(F_l(P(j\omega), K_{y0})) = 1\,(0\,dB)$, $\forall\omega$.

The solution provided by `hinfmix` (on problem 3.1) is denoted by K_{y1}. Exactly like in the one dof case, K_{y1} is, in fact, the DC-gain of the dynamic full-order controller provided by `hinfmix` and it can be checked that such a drastic reduction has no consequence on H_∞ and H_2 indexes. The static output feedback provided by `hifoo` is denoted by K_{y2}. Of course, the two solutions K_{y1} and K_{y2} meet the H_∞ constraint ($\|F_l(P, K_y)\|_\infty \leq 1$). The values of the H_2 performances are summarized in Table 3.3. We can see that `hinfmix` and `hifoo` always provide a suboptimal solution. Indeed, the frequency response of $F_l(P(s), K_{y1})$ is also depicted in Figure 3.30 (dashed black line) and highlights that this output feedback is more performant than the specification in low frequency, at the price of the degradation of the H_2 performance. The solution provided by `hifoo` depends on the initialization and can be, of course, improved using K_{y0} as an initial guess.

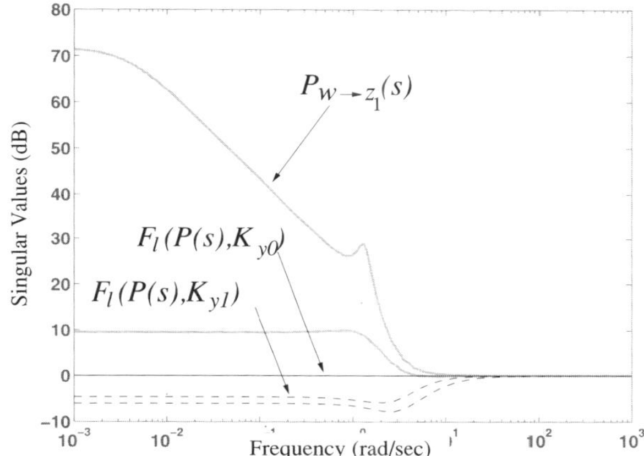

Figure 3.30. *Singular value of the open-loop transfer $P_{m_{w \to z_1}}$ and closed-loop transfer $F_l(P(s), K_y)$*

K_y	K_{y0} (nominal)	K_{y1} (`hinfmix`)	K_{y1} (`hifoo`)
$J(K_y)$	4.85	6.47	24.4

Table 3.3. *Comparison of performance indexes: $J(K_y) = \|F_l(P_m(s), K_y)_{w \to z_2}\|_2$ obtained with analytical and numerical solutions*

Note that, exactly like in the one dof case, the nominal static output feedback K_{y0} can be recovered on the pure H_∞ problem $P(s)$ (that is forgetting the H_2 channel) using:

– `hinflmi` and the DC gain of the obtained full-order controller;

– `hinfstruct` with K_{y0} as an initial guess.

From a methodological point of view, it is worth mentioning that fixed-structured H_∞ controller solvers are very efficient for designing directly a static controller but the initialization is a key element which cannot be done randomly. In that sense, reverse engineering can be very useful to provide a well-conditioned initialization, in particular when the basic problem $P(s)$ will be augmented by additional specifications or new dynamic subsystems (see the next section).

3.6.3. *Augmented H_∞ control problem*

Following what was proposed in section 3.4.3, actuators are now taken into account. It is assumed that the actuator can be modeled by a second-order transfer function with a bandwidth of 10 rad/s and a damping ratio of 0.7 on both the ailerons and rudder servo-mechanisms:

$$A(s) = \frac{100}{s^2 + 14s + 100}.$$

The new model $P_{A/C,A}(s)$ of the aircraft and the new standard problem $P_A(s)$ weighting the acceleration sensitivity function are depicted in Figure 3.31.

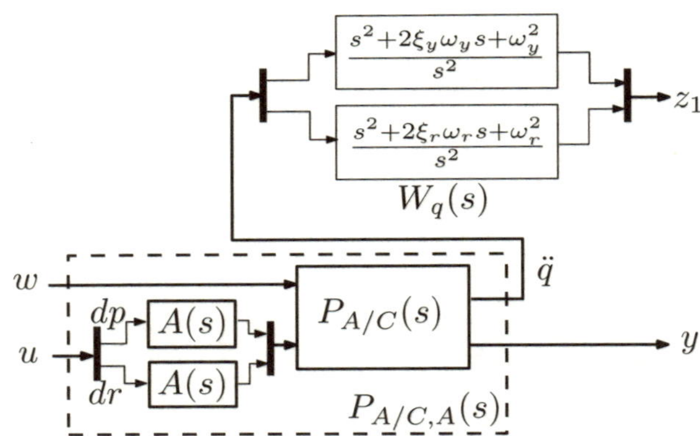

Figure 3.31. *The model $P_{A/C,A}(s)$ taking into account actuator dynamics $A(s)$ and the H_∞ standard problem $P_A(s)$ on the acceleration sensitivity function weighted by $W_q(s)$*

Note that the frequency gap between actuator bandwidth (10 rad/s) and the closed-loop bandwidth required on the yaw and roll servo-loops ($\omega_y = \omega_r = 3$ rad/s) is not wide enough to neglect actuator dynamics in the control design. Indeed, Figure 3.32 represents the step responses of:

– $F_l(P_{A/C}(s), 0_{2\times 4})$, that is the open-loop acceleration sensitivity function between w and $\ddot{q}$: we can note strong oscillations and dynamic couplings between yaw and roll.

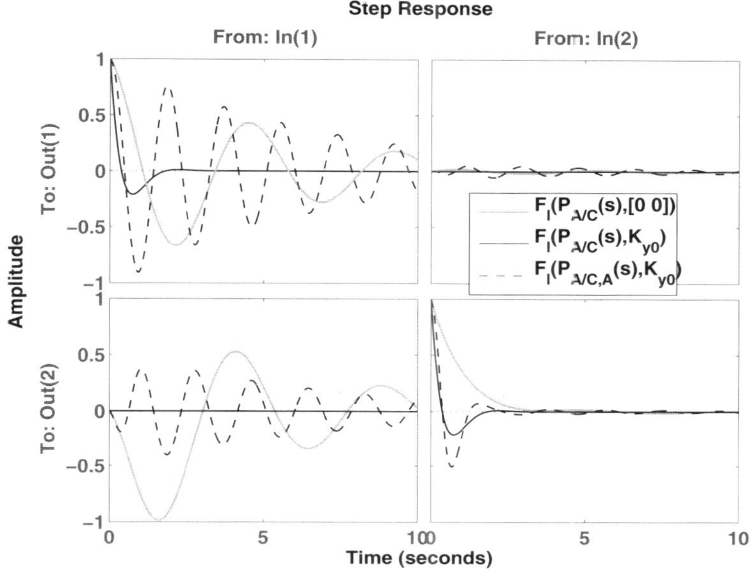

Figure 3.32. *Step responses of the acceleration sensitivity function of the A/C in open-loop and closed-loop on a static output feedback*

– $F_l(P_{A/C}(s), K_{y0})$: the nominal control law K_{y0} allows us to damp the oscillations and to decouple perfectly the roll and the yaw considering the aircraft model without actuators.

– $F_l(P_{A/C,A}(s), K_{y0})$: the previous performance is completely degraded when the actuator dynamics is taken into account. Indeed:

$$\|F_l(P_{A/C,A}(s), K_{y0})\|_\infty = 15.85$$

that is the disturbance rejection index of the nominal static output feedback K_{y0} on the model augmented with actuator dynamics is very poor and far from the specification.

3.6.3.1. *Design tacking account actuator dynamics*

To improve the closed-loop disturbance rejection ratio, a second-order controller can be designed on the H_∞ standard problem $P_A(s)$ using a fixed-structure H_∞ solver `hinfstruct`. To guarantee the yaw/roll decoupling (with an error less than 1%) and the template $W_q(s)^{-1}$ is met independently on the roll and yaw channels, the problem which is actually solved is a multichannel H_∞ problem: find a stabilizing output feedback $K_{y,dyn_1}(s)$ such that:

$$K_{y,dyn_1}(s) = \arg \min_{K_{2nd}(s)} \max \; \big(\|F_l(P_{A/C,A}(s), K_{2nd}(s))(1,1)\|_\infty, \\ \|F_l(P_{A/C,A}(s), K_{2nd}(s))(2,2)\|_\infty, \\ \|100\, F_l(P_{A/C,A}(s), K_{2nd}(s))(1,2)\|_\infty, \\ \|100\, F_l(P_{A/C,A}(s), K_{2nd}(s))(2,1)\|_\infty\big),$$

where $K_{2nd}(s)$ is in the set of second-order (2×4) controllers.

The obtained H_∞ performance index is 1.64 and highlights that the specifications are almost met (in comparison with the value 15.85 obtained with K_{y0}). Indeed, the step response of $F_l(P_{A/C,A}(s), K_{y,dyn_1}(s))$ is presented in Figure 3.33 (dashed black line): the roll/yaw decoupling is quite good and oscillations are correctly damped. But the frequency response of such a controller ($K_{y,dyn_1}(s)$), presented in Figure 3.34 (dashed black line), reveals high magnitude in high frequency (mainly around 80 rad/s). Such a behavior can be critical for actuator health.

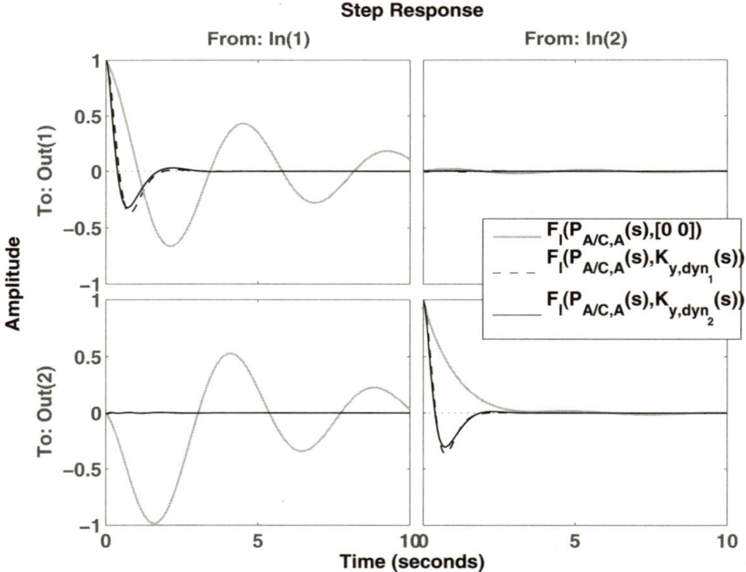

Figure 3.33. *Step responses of the acceleration sensitivity function of the A/C in open- and closed-loop on a dynamic output feedback*

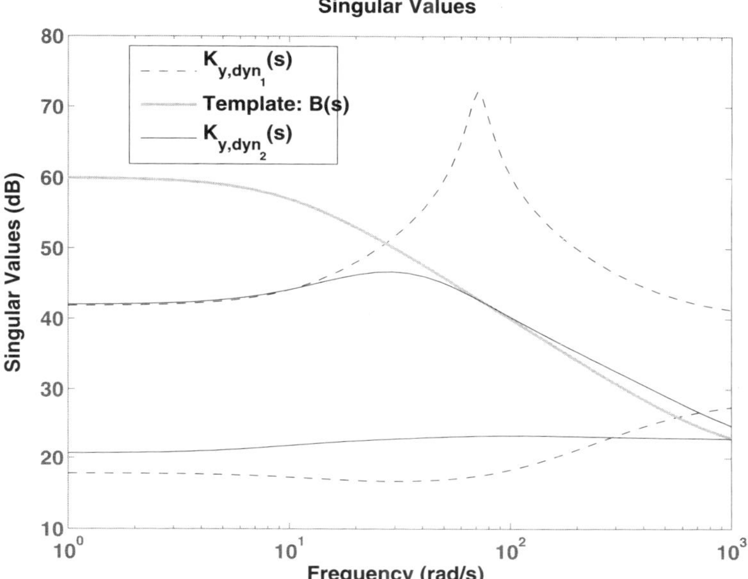

Figure 3.34. *Singular value responses of dynamic output feedbacks $K_{y,dyn_1}(s)$ and $K_{y,dyn_2}(s)$ and roll-off template $B(s)$*

3.6.3.2. *Design tacking account a roll-off specification*

To reduce the magnitude of the control law in high frequencies, a roll-off behavior is specified throughout the template:

$$B(s) = \frac{s + 1000}{0.1s + 1}$$

on the controller frequency-domain response. This template is represented in Figure 3.34 (solid grey line).

To meet this new specification, the proposed controller structure $K_{struc}(s)$ consists of the previous (2×4) $K_{2nd}(s)$ controller in a series of two single-input single-output filters $K_{dp}(s)$ and $K_{dr}(s)$ on the ailerons and the rudder, respectively, according to Figure 3.35.

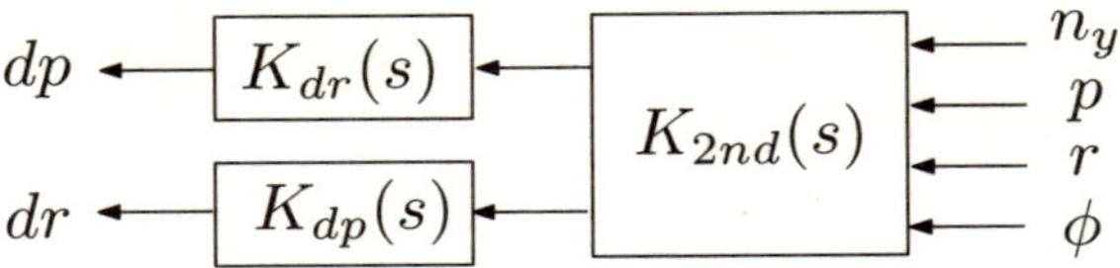

Figure 3.35. *Control law structure* $K_{struc}(s)$

The problem is thus to find a stabilizing output feedback $K_{y,dyn_2}(s)$ such that:

$$\begin{aligned} K_{y,dyn_2}(s) = \arg \min_{K_{stuct}(s)} \max \; & (\|F_l(P_{A/C,A}(s), K_{struct}(s))(1,1)\|_\infty, \\ & \|F_l(P_{A/C,A}(s), K_{struct}(s))(2,2)\|_\infty, \\ & \|100\, F_l(P_{A/C,A}(s), K_{struct}(s))(1,2)\|_\infty, \\ & \|100\, F_l(P_{A/C,A}(s), K_{struct}(s))(2,1)\|_\infty, \\ & \left\| \begin{bmatrix} B(s) & 0 \\ 0 & B(s) \end{bmatrix} K_{struct}(s) \right\|_\infty) . \end{aligned}$$

The obtained H_∞ performance index is 1.42 and highlights that the specifications are actually met. Indeed, the step response of $F_l(P_{A/C,A}(s), K_{y,dyn_2}(s))$ is presented in Figure 3.33 (solid black line): the roll/yaw decoupling is quite good, oscillations are correctly damped and undershoots on the response diagonal terms are smaller than the ones obtained with controller $K_{y,dyn_1}(s)$. The response is quite close to the one obtained with the nominal output feedback K_{y0} on the plant $P_{A/C}$ without actuator dynamics (see Figure 3.32). The frequency response of the controller ($K_{y,dyn_2}(s)$) is presented in Figure 3.34 (solid black line) and highlights that the template is almost fit according to the obtained value of the H_∞ performance.

The reader can download from http://personnel.isae.fr/daniel-alazard/matlab-packages a Matlab® script file `demo_section_3_6.m` to run all the numerical analyses presented in this section.

3.7. Conclusions

This chapter presented a new H_∞ control problem based on the acceleration sensitivity function for general mechanical systems. This scheme allows us to take into account directly basic specifications in terms of degrees-of-freedom decoupling and prescribed dynamics. It was shown how a nominal state-feedback meeting these specifications can be recovered using standard H_∞ synthesis numerical solvers and how this scheme can be used to design an improved controller to face a weak actuator dynamics while taking into account roll-off and strong stabilization specifications. One of its main advantages is certainly the very simple way to take into account specifications in the weighting system which is independent of the available measurements. Such properties were illustrated on a very simple example and on a multi-variable example: the design a lateral flight control law for an aircraft with yaw-roll decoupling specification. The reader could find in [LOQ 12] an application of this approach to the design of a three-axis attitude control system for a highly flexible earth-observation spacecraft.

The academic one degree-of-freedom mechanical system was particularly detailed to derive, analytically, the state-feedback meeting the acceleration disturbance rejection template while minimizing an H_2 performance index on the consumption. This very simple example allows us also to highlight that today there are no numerical tools to solve correctly this problem which is still open. Various cases were identified according to the frequency response of the open-loop acceleration sensitivity function w.r.t. the template. The optimal state-feedback can be recovered only when the specification requires us to increase the disturbance rejection on the whole frequency range.

The various illustrations show also the interest of recent numerical solvers for fixed-structure controller H_∞ synthesis to provide directly a low-order controller. Reverse engineering, that is the *a priori* knowledge on the controller, could be very promising to initialize such solvers which are sensitive to the initial guess.

3.8. Bibliography

[ALA 02] ALAZARD D., "Robust H_2 design for lateral flight control of a highly flexible aircraft", *Journal of Guidance, Control and Dynamics*, vol. 25, no. 3, pp. 502–509, 2002.

[ALA 08] ALAZARD D., FEZANS N., IMBERT N., "Mixed H_2/H_∞ control design for mechanical systems: analytical and numerical developments", *AIAA Guidance Navigation and Control Conference*, Honolulu, Hawaii, August 2008.

[APK 95] APKARIAN P., GAHINET P., "A convex characterization of gain-scheduled H_∞ controllers", *IEEE Transactions on Automatic Control*, vol. 40, no. 5, pp. 853–864, May 1995.

[ARZ 11] ARZELIER D., DEACONU G., GUMUSSOY S., HENRION D., "H2 for HIFOO", *International Conference on Control and Optimization with Industrial Applications*, Ankara, Turkey, August 2011.

[BOY 03] BOYD S., VANDERBERGHE L., *Convex Optimization*, Cambridge University Press, 2003.

[CHA 89] CHAMPETIER C., MAGNI J., "Analysis and synthesis of modal control laws", *La Recherche Aérospatiale*, no. 6, pp. 17–35, 1989.

[FEI 11] FEINTUCH A., "On strong stabilization of asymptotically time-invariant linear time-varying systems", *Mathematics of Control, Signals, and Systems*, vol. 22, pp. 229–243, 2011.

[FEZ 07] FEZANS N., ALAZARD D., IMBERT N., CARPENTIER B., "H_∞ control design for multivariable mechanical systems - application to RLV reentry", *AIAA Guidance Navigation and Control Conference*, Hilton Head, August 2007.

[FEZ 08] FEZANS N., ALAZARD D., IMBERT N., CARPENTIER B., "H_∞ control design for generalized second order systems based on acceleration sensitivity function", *Proceedings of the 16th Mediterranean Conference on Control and Automation*, Ajaccio-Corsica, 25–27, June 2008.

[FEZ 10] FEZANS N., ALAZARD D., IMBERT N., CARPENTIER B., "Robust LPV control design for a RLV during reentry", *AIAA Guidance Navigation and Control Conference*, Toronto, CA, August 2010.

[GAD 06] GADEWADIKAR J., LEWIS F., XIE L., KUCERA V., ABU-KHALAF M., "Parameterization of all stabilizing H_∞ static state-feedback gains: application to output-feedback design", *Proceedings of the 45th IEEE Conference on Decision and Control*, San Diego, CA, pp. 3566–3571, December 2006.

[GAH 94] GAHINET P., NEMIROVSKI A., LAMB A.J., CHILALI M., *LMI Control Toolbox*, The Mathworks Inc, 1994.

[GAU 06] GAULOCHER S., CHRÉTIEN J., PITET C., "Six-axis control design and controller switching for spacecraft formation flying", *Proceedings of the 2006 AIAA Guidance, Navigation, and Control Conference and Exhibit*, Keystone, CO, AIAA, pp. 1–20, August 2006.

[GUM 09] GUMUSSOY S., HENRION D., MILLSTONE M., OVERTON M., "Multiobjective robust control with HIFOO 2.0", *Proceedings of the IFAC Symposium on Robust Control Design*, Haïfa, Israel, June 2009.

[HIC 96] HICKS K.L., RODRIGUEZ A.A., "Decoupling compensation for the Apache helicopter", *Conference on Decision and Control*, vol. 2, Kobe, Japan, pp. 1551–1556, December 1996.

[JUN 90] JUNKINS J., *Mechanics and Control of Large Flexible Structures*, Progress in Astronautics and Aeronautics, 1990.

[LOQ 12] LOQUEN T., DE PLINVAL H., CUMER C., ALAZARD D., "Attitude control of satellites with flexible appendages: a structured H_∞ control design", *AIAA Guidance Navigation and Control Conference*, Mineapolis, MN, August 2012.

[MAG 02] MAGNI J.-F., *Robust Modal Control with a Toolbox for Use with MATLAB*, Springer, 2002.

[MEI 80] MEIROVITCH L., *Computational Methods in Structural Dynamics*, Sijthoff and Noordhoff, 1980.

[SUZ 06] SUZUKI R., TANI M., IKEMOTO M., FUJIKI N., KOBAYASHI N., HOFER E.P., "Decoupling property of full information H_∞ control and its application to mechanical systems", *International Conference on Control Applications*, Munich, Germany, pp. 734–739, October 2006.

[TIS 96] TISCHLER M.B., *Advances in Aircraft Flight Control*, Taylor & Francis, 1996.

[VOI 03] VOINOT O., ALAZARD D., APKARIAN P., MAUFFREY S., CLÉMENT B., "Launcher attitude control: discrete-time robust design and gain-scheduling", *Control Engineering Practice*, vol. 11, pp. 1243–1252, 2003.

[ZHO 96] ZHOU K., DOYLE J.C., GLOVER K., *Robust and Optimal Control*, Prentice Hall, 1996.

Conclusions and Perspectives

The objective of this book was to give a first insight into the field of reverse engineering for control design. The motivation is to fill the gap between the panel of modern optimal control tools available today and their practical application with a critical approach. On a simple control design problem, if control synthesis tools are not able to provide a simple controller that could be designed by an experienced control engineer, then such tools are not satisfactory from a methodological point of view and some works have to be performed to demonstrate that these tools can reproduce simple solutions before demonstrating their capabilities to solve more complex problems where practical approaches are limited. This way of thinking is required to facilitate the transfer of new control design techniques in the practical world. With these considerations in mind, this book has focused on how to take into account a given initial controller in the general H_∞ and H_2 synthesis framework. Classical H_∞ synthesis approaches (for instance, mixed-sensitivity approach) are based on a weighting system acting on the input and output signals of the plant. We think that such an approach is very limited and cannot be used to provide a simple good-sense solution. One of a the reasons is that they are based on convex optimization algorithms that produce full-order controllers. There is, of course, the guarantee to find the global optimum but at the price of conservative relaxations or sufficient conditions and an over-parameterized controller. The way to manage all controller tuning parameters (or decision variables) is very complex and could lead to a marginal solution when full-order H_∞ control is applied to a simple problem (the example proposed in Appendix 1 was quite illustrative of this problem). To overcome these problems and to develop control schemes allowing an initial controller to be taken into account, the H_∞ control schemes proposed in this book take advantage of the internal structure of the plant and use a weighting system acting:

– directly on the state equation in the case of the so-called cross standard form. The cross standard form is a general solution to the inverse H_∞ and H_2 optimal control problems. Some illustrations were proposed to highlight how a given initial control can be improved to face new frequency-domain specifications or system updates, like the actuator and sensor dynamics.

– directly on the acceleration for the particular case of mechanical systems. The acceleration sensitivity function-based H_∞ control scheme allows us to handle directly the basic specifications in terms of degrees-of-freedom decoupling and prescribed dynamics. The weighting function is independent of the used measurements. In the same spirit of the cross standard form, the standard problem, based on the acceleration sensitivity function, can be used and augmented to take into account additional specifications in order to develop a complete methodology that can be adapted for practical control design problems in the field of mechanical systems.

The cross standard form is based on the observer-based realization of a given controller and a given plant. This topic was analyzed in detail in Chapter 1 because it gives practical tools to understand controller internal state variables and to implement this controller with additional objectives: plant state and disturbance monitoring, gain scheduling and input reference tracking when the initial controller was designed on a pure disturbance rejection problem. Indeed, a servo-loop system has to satisfy several objectives. The first objective that comes to mind is the reference input tracking, but we think that disturbance rejection is a more challenging objective because disturbance is an unknown signal whereas the reference input is a known signal and it is always possible to add a feedforward controller to improve the response to a reference input. The observer-based controller, as highlighted in the illustrations of section 1.9, is quite useful to plug judiciously the reference input into the feedback loop.

It was also shown that the recent fixed-structure H_∞ synthesis methods overcome the methodological problems encountered with full-order approaches. In that sense, they are powerful tools with narrow links with the reverse engineering approach proposed in this book, for instance in the initialization (initial guess) required in the design procedure of fixed-structure controllers. Such an initialization cannot be done randomly so that the approach will be fully satisfactory from a methodological point of view.

Although fixed-structure H_∞ control design is quite advantageous, there are still some open problems:

– more particularly in the field of H_2/H_∞ control design: the proposed benchmark in section 3.5.3 was very simple and quite illustrative;

– more generally, one of the most important challenges for future research is certainly the development of tools that are able to identify the appropriate control structure when the practical engineering know-how fails to propose an initial controller due to the complexity of the problem.

APPENDICES

Appendix 1

A Preliminary Methodological Example

The objective of this preliminary example is to highlight the methodological short-comings of the classical full-order H_∞ synthesis approach in comparison with classical engineering approach and the importance of recent fixed-structure H_∞ synthesis approaches [GAH 11, ARZ 11, GUM 09] to encounter these short-comings.

Let us consider the quite simple mechanical system shown in Figure A1.1. A pneumatic cylinder can apply a force u on a mass m through a flexible transmission characterized by a stiffness k and a viscous friction coefficient d.

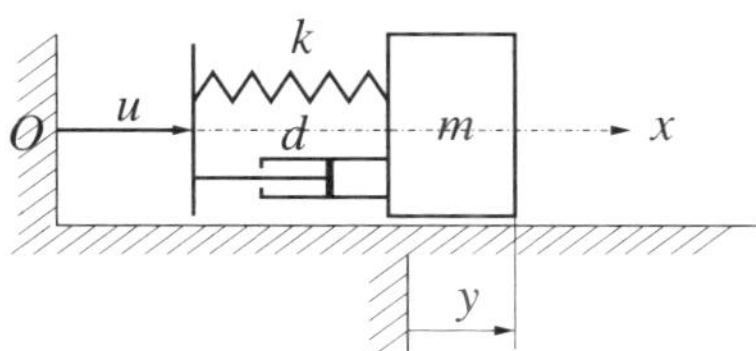

Figure A1.1. *A simple mechanical system*

Numerical application: $m = 1\,(\text{Kg})$, $d = 0.01\,(\text{Ns/m})$, $k = 1\,(\text{N/m})$.

In the above figure, y is the position of the mass relative to the rest position along the cylinder axis $(0, x)$. The objective is to servo-loop the output y to a reference input r with the specifications:

S1: servo-loop bandwidth $\omega_{bd} = 10\,\text{rad/s}$,

S2: static error on the step response $\epsilon_\infty <= 0.01$.

The model $G(s)$ between the input u and the output y reads: $G(s) = \frac{1}{ms^2+ds+k}$. The numerical application leads to:

$$G(s) = \frac{1}{s^2 + 0.01s + 1},$$

that is a poorly damped second-order system.

The specifications can be easily handled using a weight function $W_1(s)$ on the output sensitivity function $S\left(K(s)\right) = (I + G(s)K(s))^{-1}$. Considering:

$$W_1(s) = \frac{s+10}{s+0.1},$$

the frequency-domain response (magnitude) of the template $1/W_1(s)$ to be satisfied by the output sensitivity function S is shown in Figure A1.3 (solid grey line) and highlights a -40 dB magnitude in low frequency to meet specification S2 and a 0 dB magnitude beyond the pulsation 10 rad/s according to specification S1.

The standard H_∞ problem $P_1(s)$ to minimize $\|W_1(s)S\left(K(s)\right)\|_\infty$ is shown in Figure A1.2.

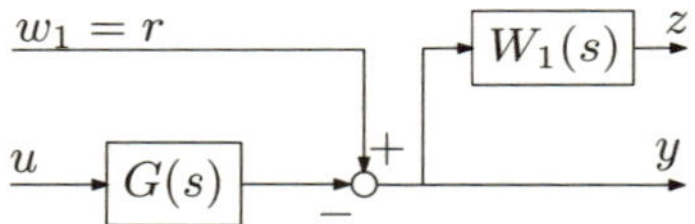

Figure A1.2. *Standard H_∞ problem $P_1(s)$ / $F_l(P_1, K) = W_1/(1 + GK) = W_1S(K)$*

The direct application of the full-order H_∞ synthesis (Matlab® function `hinfsyn`) provides a third-order controller $K_1(s)$ and the performance index:

$$\gamma_1 = \|F_l(P_1(s), K_1(s)\|_\infty = 1.0004.$$

This value is very close to one and we can conclude that the obtained sensitivity function satisfies the template $1/W_1(s)$. Indeed, the frequency-domain response of $1/(1 + G(s)K_1(s))$, shown in Figure A1.3 (dashed black line), is superposed on the template. Therefore, the specifications are satisfied. But an in-depth analysis of controller $K_1(s)$:

$$K_1(s) = \frac{100(s^2 + 0.01s + 1)}{10s + 1} \frac{1}{(8.3112\,10^{-11}s^2 + 1.3040\,10^{-5}s + 1.0106)}$$

reveals that $K_1(s)$ cancels the poles of the plant $G(s)$ with a pair of auto-conjugate zeros. In other words, the behavior of the controller on a wide frequency range reads:

$$K_1(s) \approx \frac{100}{10s + 1} G^{-1}(s).$$

It is well-known that the inversion of the model $G(s)$ in the controller must be avoided. Otherwise, the servo-loop system will be very sensitive to uncertainties on parameters m, d and k or, even, it will not reject external disturbances. Indeed, the impulse response of y to a plant input disturbance d $(G/(1 + GK_1))$, presented in Figure A1.4, reveals strong oscillations at the plant natural frequency 1 rad/s. Therefore, controller $K_1(s)$ does not at all reject external disturbances and we can conclude that $K_1(s)$ is not a servo-loop controller.

Classical control engineering allows us to state that a basic proportional-derivative (PD) controller:

$$K_2(s) = K_p + \frac{K_d s}{1 + \tau s}$$

is quite efficient to meet the specifications. The three parameters K_p, K_d and τ can be easily tuned considering the following constraints:

– $K_p G(0) \geq (1/\epsilon_\infty - 1)$ to satisfy S2, where $G(0)$ is the DC gain of $G(s)$, for instance $K_p = 100$ N/m;

– K_d must be high enough to damp the closed-loop dynamics that is governed by the characteristic polynomial $s^2 + (d + K_d)s + (k + K_p)$ (if $\tau = 0$), for instance $K_v = 15$ Ns/m,

– $1/\tau \gg \omega_{bd}$, for instance $\tau = 0.001$ s.

Then:

$$K_2(s) = \frac{15.1s + 100}{0.001s + 1}.$$

Note that the H_∞ performance of $K_2(s)$ on the initial H_∞ problem $P_1(s)$ is very close to the optimal value:

$$\gamma_2 = \|F_l(P_1(s), K_2(s))\|_\infty = 1.0643.$$

The frequency-domain response of the output sensitivity function $1/(1 + GK_2)$ and the impulse response of $G/(1 + GK_2)$ are also shown in Figures A1.3 and A1.4, respectively:

– The output sensitivity function obtained with $K_2(s)$ reveals that, between 0.1 and 10 rad/s, the performance is better than the template $1/W_1(s)$ and there is a notch around the plant natural frequency. Indeed, the output disturbance rejection is naturally efficient around the resonance frequency located inside the control bandwidth.

– The controller K_2 is efficient to reject input disturbances (Figure A1.4).

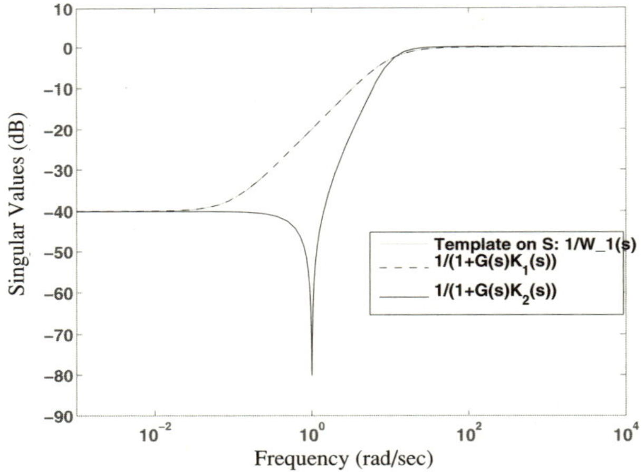

Figure A1.3. *Frequency-domain responses of output sensitivity functions and template* $1/W_1(s)$

The problems encountered with controller $K_1(s)$ are, in fact, due to the over parameterization of the controller required to convexify the optimization problem and to find the global optimum (using Matlab® function `hinfsyn`, for instance). Fixing the structure of the controller to a first-order controller overcomes all these problems. Today, fixed-structure H_∞ synthesis tools are available and it can be easy to check that, on the initial problem $P_1(s)$, functions `hinfstruct` and `hifoo` provide first-order controllers $K_3(s)$ and $K_4(s)$, respectively:

$$K_3(s) = \frac{17.2615s + 98.7862}{1.317\,10^{-4}s + 1} \text{ and } \gamma_3 = \|F_l(P_1(s), K_3(s))\|_\infty = 1.0023,$$

$$K_3(s) = \frac{18.1946s + 102.7}{7.880\,10^{-4}s + 1} \text{ and } \gamma_4 = \|F_l(P_1(s), K_4(s))\|_\infty = 1.014\,.$$

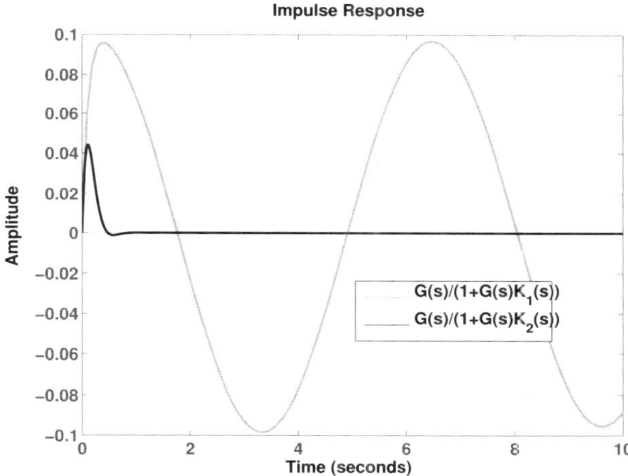

Figure A1.4. *Impulse responses $y(t)$ to an input disturbance $d(t)$*

Both controllers have a very good performance index and have the same properties as the controller $K_2(s)$, particularly regarding the input disturbance rejection. Thus, although these solvers only guarantee to converge to a local minimum, they provide a quite efficient solution while the global optimum provided by full-order H_∞ solver converges to a marginal solution. Note also that looking for a suboptimal full-order controller does not solve this problem: the pole/zeros cancelation is still present even for a suboptimal controller.

The way to manage all the decision variables in the full-order H_∞ synthesis to avoid such a marginal solution is quite complex and requires us to express additional specifications that were implicit using the classical engineering background. In the proposed example, the reader could argue that the full-order H_∞ designs have found the optimal solution for the given performance index (here $\min_K \|W_1(s)S(K(s))\|_\infty$) and that it is not quite fair to evaluate the solution in response to an input disturbance since such a disturbance was not taken into account in the H_∞ problem $P_1(s)$. Considering the problem $P_2(s)$ shown in Figure A1.5 where the input disturbance d is taken into account through the weight W_2, the full-order H_∞ design allows the pole/zero cancellation to be avoided more or less according to the weight W_2. The problem is: how $W_2(s)$ can be tuned to find a good trade-off? What is the specification on the transfer from the input disturbance to the servo-loop error? This specification is implicit using a classical PD control because there is a direct relationship between the reference input tracking performance S and the input disturbance rejection performance $-SG$. This relationship is $S + SGK = I$ and depends on the unknown controller K. So an *a priori* knowledge of K is required to specify a pertinent shape for W_2. This is one of the first arguments to develop reverse engineering approaches.

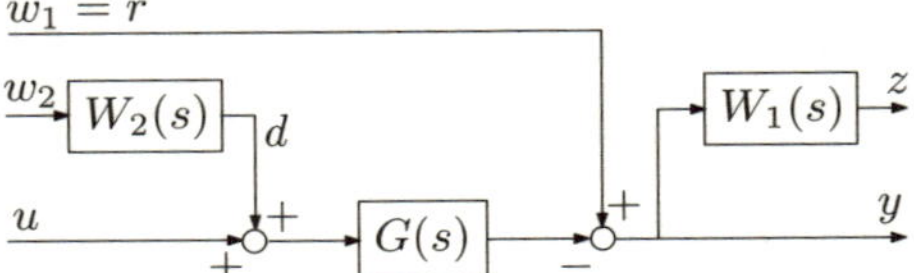

Figure A1.5. *Standard H_∞ problem*
$P_2(s)\,/\,F_l(P_2(s), K(s)) = [W_1S \;\; W_1SGW_2]$

An other important drawback is the adaptation of the method based on the output sensitivity function S when a new measurement is available; here, the velocity $\dot{y}$. Considering the problem $P_3(s)$ depicted in Figure A1.6 as the direct adaptation of the problem $P_1(s)$ with two measurements y and $\dot{y}$, it can be easy to check that `hinfsyn` produce a 1×2 third-order controller that reads:

$$U = K_1(s)Y + 0\dot{Y}.$$

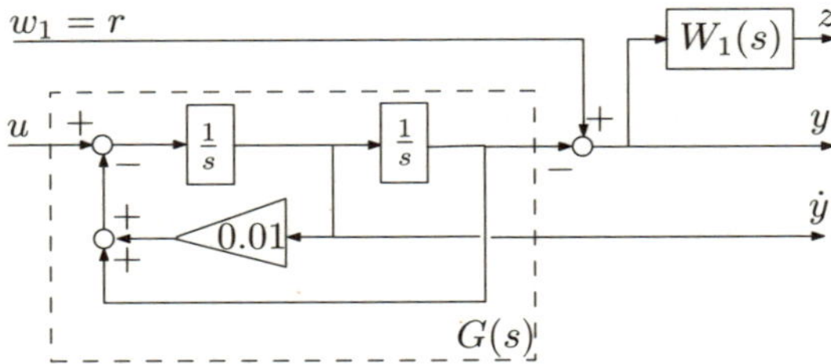

Figure A1.6. *Standard H_∞ problem $P_3(s)$: position sensitivity function with two measurements*

That is, the second measurement $\dot{y}$ is not used. In fact, while it is obvious that a pair of PD gains is sufficient to meet the specifications on the position sensitivity function, the adaptation of the H_∞ control problem $P_1(s)$ to the two outputs plant requires a 2×2 weighting function $W_1(s)$. This weight must take into account that the velocity is the time derivative of the position and is not at all easy to be expressed. Note that imposing a static controller in fixed-structure H_∞ synthesis (`hinfstruct` or `\hifoo`) allows us to find that a pair of gains can easily meet the specifications on the problem $P_3(s)$.

The reader could also argue that this example is too simple to highlight the importance of optimal H_∞ control design, but from a practical point of view, control engineers need to be confident of a new synthesis method and for this purpose, they must understand how this method can provide a well-known solution before providing a better solution. So, there is an actual need to develop new H_∞ control methods to link classical engineering approaches and optimal approaches. Thus, when the classical engineering approach fails to find a good tuning due to the problem of complexity, such a new method can provide some solutions in which we can be confident.

The reader can upload, from the web page:
`http://personnel.isae.fr/daniel-alazard/matlab-packages`,
a Matlab® script file `demo_appendix_1.m` to run all the numerical analyses presented in this appendix.

A1.1. Bibliography

[ARZ 11] ARZELIER D., DEACONU G., GUMUSSOY S., HENRION D., "H2 for HIFOO", *International Conference on Control and Optimization with Industrial Applications*, Ankara, Turkey, August 2011.

[GAH 11] GAHINET P., APKARIAN P., "Structured H_∞ synthesis in MATLAB", *Proceedings of the 18th IFAC World Congress*, Milan, Italy, pp. 1435–1440, 28 August–2 September 2011.

[GUM 09] GUMUSSOY S., HENRION D., MILLSTONE M., OVERTON M., "Multiobjective robust control with HIFOO 2.0", *Proceedings of the IFAC Symposium on Robust Control Design*, Haïfa, Israel, June 2009.

Appendix 2

Discrete-time Case

Techniques presented in Chapters 1 and 2 in the continuous-time case are now extended to the discrete-time case (proofs are omitted for brevity).

The discrete-time plant $G(z)$ (order n) is defined as:

$$\begin{cases} x(k+1) = Ax(k) + Bu(k) \\ y(k) \quad\;\;\; = Cx(k) + Du(k) \end{cases} \qquad [A2.1]$$

The discrete-time controller $K_0(z)$ (order n_K) is defined as:

$$\begin{cases} x_K(k+1) = A_K x_K(k) + B_K y(k) \\ u(k) \qquad\;\;\; = C_K x_K(k) + D_K u(k) \end{cases} \qquad [A2.2]$$

Two classical implementation structures of discrete-time observer-based controllers can be used: the predictor and the estimator structures.

A2.1. Discrete-time predictor form

The predictor form is described by:

$$\begin{cases} \widehat{x}(k/k) \qquad\; = A\widehat{x}(k/k-1) + Bu(k) & \text{Prediction} \\ \widehat{x}(k+1/k) = \widehat{x}(k/k) + K_f(y(k) - C\widehat{x}(k/k-1) - Du(k)) & \text{Correction} \\ u(k+1) \qquad = -K_c\widehat{x}(k+1/k) & \text{Control} \end{cases} \qquad [A2.3]$$

This case is analogous to the continuous-time case. The construction procedure is, therefore, the same. It provides the parameters K_c^p, K_f^p, A_Q^p, B_Q^p, C_Q^p and D_Q^p of the

Youla parameterization associated with the predictor form whose state-space representation reads:

$$\begin{cases} \widehat{x}(k+1/k) &= A\widehat{x}(k/k-1) + Bu(k) + K_f^p(y(k) - C\widehat{x}(k/k-1) - Du(k)) \\ x_Q(k+1) &= A_Q^p x_Q(k) + B_Q^p(y(k) - C\widehat{x}(k/k-1) - Du(k)) \\ u(k) &= -K_c^p\widehat{x}(k/k-1) + C_Q^p x_Q(k) + D_Q^p(y(k) - C\widehat{x}(k/k-1) - Du(k)) \end{cases} \quad \text{[A2.4]}$$

A2.2. Discrete-time estimator form

The estimator structure of an observer-based controller is now described as:

$$\begin{cases} \widehat{x}(k+1/k) &= A\widehat{x}(k/k) + Bu(k) & \text{Prediction} \\ \widehat{x}(k+1/k+1) &= \widehat{x}(k+1/k) + K_f(y(k+1) & \\ & \quad -C\widehat{x}(k+1/k) - Du(k+1)) & \text{Correction} \\ u(k+1) &= -K_c\widehat{x}(k+1/k+1) & \text{Control} \end{cases} \quad \text{[A2.5]}$$

In contrast to the previous case, this discrete-time estimator controller exhibits a direct feed-through between $y(k)$ and $u(k)$ but the separation principle still holds: the closed-loop transfer function between the reference input and the innovation $y(k) - C\widehat{x}(k/k-1) - Du(k)$ is zero and the closed-loop poles can be split into the closed-loop state-feedback poles ($\mathrm{spec}(A - BK_c)$), which are unobservable from the innovation, and the closed-loop state-estimator poles ($\mathrm{spec}(A(I - K_fC))$), which are uncontrollable by the reference input. The Youla parameterization associated with this structure is depicted in Figure A2.1 and reads:

$$\begin{cases} \widehat{x}(k+1/k) &= A\widehat{x}(k/k-1) + Bu(k) + AK_f(y(k) - C\widehat{x}(k/k-1) - Du(k)) \\ x_Q(k+1) &= A_Q x_Q(k) + B_Q(y(k) - C\widehat{x}(k/k-1) - Du(k)) \\ u(k) &= -K_c\widehat{x}(k/k-1) + C_Q x_Q(k) \\ & \quad +(D_Q - K_cK_f)(y(k) - C\widehat{x}(k/k-1) - Du(k)) \end{cases} \quad \text{[A2.6]}$$

We know from sections 1.4 and A2.1 how to compute all the parameters (K_c^p, K_f^p, A_Q^p, B_Q^p, C_Q^p and D_Q^p) of the predictor form and the corresponding Youla parameterization, from a given controller (A_K, B_K, C_K, D_K) and a given plant (A, B, C, D). As a result, the parameters (K_c, K_f, A_Q, B_Q, C_Q and D_Q) of the equivalent estimator form can be obtained by direct identification of the representations [A2.4] and [A2.6]. This yields:

$$\begin{aligned} &K_c = K_c^p, \quad K_f = A^{-1}K_f^p, \\ &A_Q = A_Q^p, \quad B_Q = B_Q^p, \quad C_Q = C_Q^p, \quad D_Q = D_Q^p + K_c^pK_f^p \end{aligned} \quad \text{[A2.7]}$$

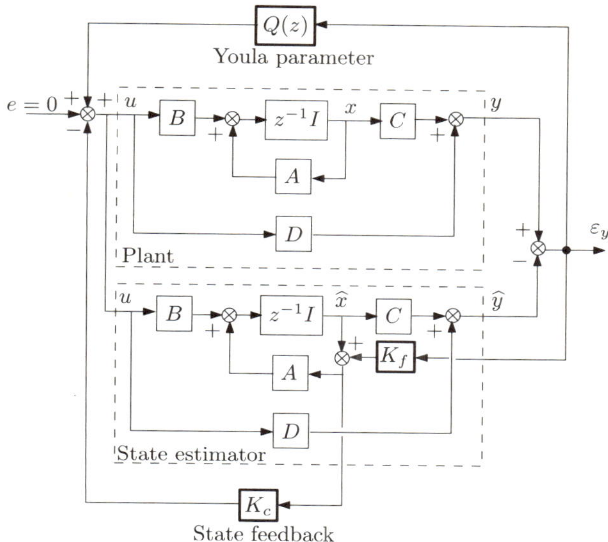

Figure A2.1. *The discrete-time Youla parameterization using state estimator structure (where* $\widehat{x}_k = \widehat{x}(k/k-1)$*)*

A2.3. Discrete-time cross standard form

In the case of a low-order controller ($n_K \leq n$), the general expression for the CSF (equation [2.2]) is valid for the discrete-time case.

In the case of the augmented-order controller ($n_K \geq n$), it is possible to define the CSF associated with an estimator form of the controller [A2.6], This CSF reads:

$$P_{CSF}(z){:} = \left[\begin{array}{cc|cc} A & 0 & AK_f & B \\ 0 & A_Q & B_Q & 0 \\ \hline K_c & -C_Q & -D_Q + K_cK_f & I_m \\ C & 0 & I_p & D \end{array}\right] \qquad [A2.8]$$

The block diagram associated with this CSF is depicted in Figure A2.2.

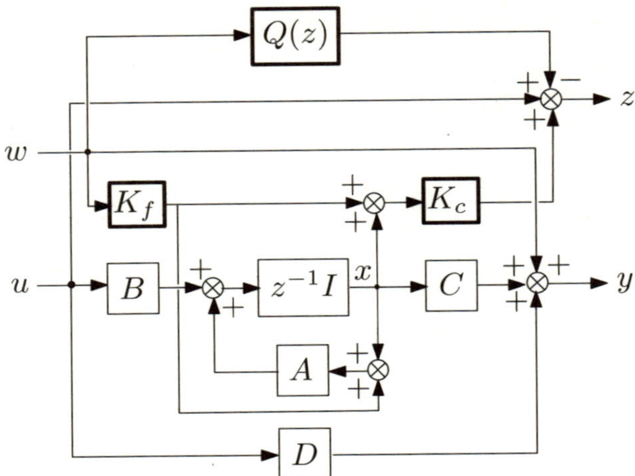

Figure A2.2. *Discrete-time cross standard form*

Appendix 3

Nominal State-feedback for Mechanical Systems

Let us consider multi-variable mechanical systems described by the generalized second-order differential equation:

$$\mathbf{M}\ddot{q} + \mathbf{D}\dot{q} + \mathbf{K}q = \mathbf{F}u. \qquad [A3.1]$$

The following proofs are restricted to the case where $\mathbf{F}$ is a square and invertible. Let $\mathbf{F}^{-1}$ be its inverse and $\mathbf{F}^{-\mathrm{T}} = (\mathbf{F}^{-1})^{\mathrm{T}} = (\mathbf{F}^{\mathrm{T}})^{-1}$.

A3.1. Recovering control law [3.4] using linear-quadratic approach

Consider the state vector $x = \begin{bmatrix} q^{\mathrm{T}} & \dot{q}^{\mathrm{T}} \end{bmatrix}^{\mathrm{T}}$ and the following state-space representation of a generalized second-order system as defined in equation [A3.1].

$$\dot{x} = \begin{bmatrix} 0 & I_n \\ -\mathbf{M}^{-1}\mathbf{K} & -\mathbf{M}^{-1}\mathbf{D} \end{bmatrix} x + \begin{bmatrix} 0 \\ \mathbf{M}^{-1}\mathbf{F} \end{bmatrix} u = Ax + Bu \qquad [A3.2]$$

Let A_d be the desired closed-loop dynamic matrix filling specifications, that is each degree of freedom q_i must have a second-order dynamics behavior defined by a pulsation ω_i and a damping ratio ξ_i and must be two-by-two decoupled:

$$A_d = \begin{bmatrix} 0 & I_n \\ -\mathrm{diag}(\omega_i) & -\mathrm{diag}(2\xi_i\omega_i) \end{bmatrix}. \qquad [A3.3]$$

Using the implicit reference model approach, let us define J as follows:

$$J = \int_0^{+\infty} (\dot{x} - A_d x)^{\mathrm{T}}(\dot{x} - A_d x)\mathrm{dt} \qquad \text{[A3.4]}$$

$$= \int_0^{+\infty} x^{\mathrm{T}} \underbrace{(A - A_d)^{\mathrm{T}}(A - A_d)}_{Q} x + u^{\mathrm{T}} \underbrace{B^{\mathrm{T}}B}_{R} u + 2x^{\mathrm{T}} \underbrace{(A - A_d)^{\mathrm{T}}B}_{N} u \,\mathrm{dt}$$

The solution minimizing J is the control law $u = -K_c\, x$ with $K_c = R^{-1}(B^{\mathrm{T}}P_c + N^{\mathrm{T}})$ and P_c is the positive solution of equation [A3.5].

$$P_c(A - BR^{-1}N^{\mathrm{T}}) + (A - BR^{-1}N^{\mathrm{T}})P_c - P_c BR^{-1}B^{\mathrm{T}}P_c + Q - NR^{-1}N^{\mathrm{T}} = 0. \qquad \text{[A3.5]}$$

Considering:

$$A - A_d = \begin{bmatrix} 0_n & 0_n \\ -\mathbf{M}^{-1}\mathbf{K} + \mathrm{diag}(\omega_i^2) & -\mathbf{M}^{-1}\mathbf{D} + \mathrm{diag}(2\xi_i\omega_i) \end{bmatrix}$$

$$\begin{aligned} R &= \begin{bmatrix} 0 \ \mathbf{F}^{\mathrm{T}}\mathbf{M}^{-\mathrm{T}} \end{bmatrix} \begin{bmatrix} 0 \\ \mathbf{M}^{-1}\mathbf{F} \end{bmatrix} = \mathbf{F}^{\mathrm{T}}\mathbf{M}^{-\mathrm{T}}\mathbf{M}^{-1}\mathbf{F} \\ R^{-1} &= \mathbf{F}^{-1}\mathbf{M}\mathbf{M}^{\mathrm{T}}\mathbf{F}^{-\mathrm{T}} \\ N^{\mathrm{T}} &= B^{\mathrm{T}}(A - A_d) \\ &= \begin{bmatrix} \mathbf{F}^{\mathrm{T}}\mathbf{M}^{-\mathrm{T}}(-\mathbf{M}^{-1}\mathbf{K} + \mathrm{diag}(\omega_i^2)) \ \mathbf{F}^{\mathrm{T}}\mathbf{M}^{-\mathrm{T}}(-\mathbf{M}^{-1}\mathbf{D} + \mathrm{diag}(2\xi_i\omega_i)) \end{bmatrix}. \end{aligned}$$

It can easily be shown that:

$$NR^{-1}N^{\mathrm{T}} = (A - A_d)^{\mathrm{T}}(A - A_d) = Q. \qquad \text{[A3.6]}$$

Using [A3.6], it is obvious that $P_c = 0$ is the solution of equation [A3.5]. As a result, the solution is:

$$K_c = R^{-1}N^{\mathrm{T}} = -\mathbf{F}^{-1}\begin{bmatrix} \mathbf{K} - \mathbf{M}\mathrm{diag}(\omega_i^2) \ \mathbf{D} - \mathbf{M}\mathrm{diag}(2\xi_i\omega_i) \end{bmatrix} = \mathbf{K_0}. \qquad \text{[A3.7]}$$

This solution is identical to the solution given by mechanical approach equation [3.4].

A3.2. Recovering control law [3.4] using eigenstructure assignment

To adapt the classical eigenstructure assignment [MAG 02] approaches to this problem, eigenvalues λ_i are defined from desired pulsations and damping ratios (ω_i, ξ_i), $i = 1,\ 2, \cdots,\ n$ as follows:

$$\forall i \in [1, n]\,, \lambda_i = -\xi_i\,\omega_i + j\,\sqrt{1-\xi_i^2}\,\omega_i \qquad [A3.8]$$

$$\forall i \in [n+1, 2n]\,, \lambda_i = \overline{\lambda}_{i-n} \qquad [A3.9]$$

For each eigenvalue λ_i, the associated eigenvector v_i is decoupled from $q_j, \forall j \neq i$. So let us define I_n^i as the $n-1 \times n$ matrix built removing the ith row of I_n and $E_i = \left[I_n^i\ 0_{n-1\times n}\right]$. Assuming A has the structure presented in equation [A3.2], then for $i \in [1, \cdots, n]$, v_i and w_i can be computed such that equation [A3.10] is true.

$$\underbrace{\begin{bmatrix} A + \lambda_i\, I_n & B \\ E_i & 0_{n-1\times n} \end{bmatrix}}_{\mathcal{A}_i} \begin{bmatrix} v_i \\ w_i \end{bmatrix} = 0 \qquad [A3.10]$$

That is $[v_i^{\mathrm{T}} \quad w_i^{\mathrm{T}}]^{\mathrm{T}} \in \mathrm{Ker}(\mathcal{A}_i)$. The corresponding state-feedback law is:

$$K = -\left[w_1 \cdots w_n\ \overline{w}_1 \cdots \overline{w}_n\right] \left[v_1 \cdots v_n\ \overline{v}_1 \cdots \overline{v}_n\right]^{-1} \qquad [A3.11]$$

Due to decoupling constraints E_i, it can be shown that $K = \mathbf{K_0}$.

A3.3. Recovering control law [3.4] using LMI

Illustrations presented in sections 3.4.2, 3.5.3 and 3.6 showed how it was possible to recover the nominal static state-feedback gain $\mathbf{K_0}$ defined in equation [3.4] using a full-order LMI-based H_∞ solver followed by a reduction of the full-order controller to its DC gain. This can be explained using the parameterization of all stabilizing state-feedbacks.

A3.3.1. *Parameterization of all stabilizing state-feedback gains*

In [ZHO 96] Therorem 17.6, and [GAD 06], parameterization of all stabilizing state-feedback with a guarantee on the closed-loop H_∞ performance is proposed. Let us consider the standard problem $P(s)$ (equation [3.9]) shown in Figure 3.3 and the change of variable $u = \mathbf{F}^{-1}\mathbf{M}\widetilde{u}$. Then the new problem $\widetilde{P}$ between z_1 and $[w^T \quad \widetilde{u}^T]^T$ writes:

$$\widetilde{P} := \left[\begin{array}{cc|c|c} 0_n & I_n & 0_n & 0_n \\ -\mathbf{M}^{-1}\mathbf{K} & -\mathbf{M}^{-1}\mathbf{D} & I_n & I_n \\ \hline \text{diag}(\omega_i^2) - \mathbf{M}^{-1}\mathbf{K} & \text{diag}(2\xi_i\omega_i) - \mathbf{M}^{-1}\mathbf{D} & I_n & I_n \\ \hline I_n & 0_n & 0_n & 0_n \\ 0_n & I_n & 0_n & 0_n \end{array}\right] = \left[\begin{array}{c|c|c} A & B_2 & B_2 \\ \hline C_1 & I & I \\ \hline I & 0 & 0 \end{array}\right].$$

Considering a static state-feedback $\widetilde{u} = \widetilde{K}x$, the closed-loop plant reads:

$$F_l(\widetilde{P}, \widetilde{K}) := \left[\begin{array}{c|c} A + B_2\widetilde{K} & B_2 \\ C_1 + \widetilde{K} & I \end{array}\right],$$

Then for a given γ, the following assertions are equivalent to characterize an admissible stabilizing state-feedback K:

– $\|F_l(\widetilde{P}, \widetilde{K})\|_\infty < \gamma$,

– $H(\gamma) = \begin{bmatrix} A + B_2\widetilde{K} + \frac{1}{\gamma^2-1}B_2(C_1+\widetilde{K}) & \frac{1}{\gamma^2-1}B_2B_2^T \\ -\frac{\gamma^2}{\gamma^2-1}(C_1+\widetilde{K})^T(C_1+\widetilde{K}) & -\left(A + B_2\widetilde{K} + \frac{1}{\gamma^2-1}B_2(C_1+\widetilde{K})\right)^T \end{bmatrix}$

has no eigenvalue on the imaginary axis.

– $\exists\, X > 0$ solution of:

$$\begin{aligned} & X\left(A + B_2\widetilde{K} + \frac{1}{\gamma^2-1}B_2(C_1+\widetilde{K})\right) \\ & + \left(A + B_2\widetilde{K} + \frac{1}{\gamma^2-1}B_2(C_1+\widetilde{K})\right)^T X \\ & + X\frac{B_2B_2^T}{\gamma^2-1}X + \frac{\gamma^2}{\gamma^2-1}(C_1+\widetilde{K})^T(C_1+\widetilde{K}) = 0, \end{aligned}$$

– or (after some developments) $\exists\, X > 0$ solution of:

$$\begin{aligned} & X(A - B_2C_1) + (A - B_2C_1)^T X - XB_2B_2^T X \\ & + \frac{\gamma^2}{\gamma^2-1}(\widetilde{K} + C_1 + B_2^T X)^T(\widetilde{K} + C_1 + B_2^T X) = 0. \end{aligned}$$

Then the parameterization reads $\widetilde{K}$ is a stabilizing state-feedback gain such that $\|F_l(\widetilde{P}, \widetilde{K})\|_\infty \leq \gamma$ if and only if there exists a parameter L such that: $\widetilde{K} = -C_1 - B_2^T X + L$, where $X = X^T \geq 0$ solution of:

$$X(A - B_2C_1) + (A - B_2C_1)^T X - XB_2B_2^T X + \frac{\gamma^2}{\gamma^2-1}L^T L = 0.$$

The central control $\widetilde{K}_0$ is defined for the particular value $L = 0$, that is:

$$\widetilde{K}_0 = -C_1 - B_2^T X_0 \text{ with } X_0 = X_0^T \geq 0 \text{ solution of:}$$

$$X_0(A - B_2C_1) + (A - B_2C_1)^T X_0 - X_0 B_2 B_2^T X_0 = 0.$$

In our problem, $A - B_2C_1 = \begin{bmatrix} 0_n & I_n \\ -\text{diag}(\omega_i^2) & -\text{diag}(2\xi_i\omega_i) \end{bmatrix}$ is a stable matrix and depends only on specifications $(\xi_i,\ \omega_i)$ and not on the plant characteristics (**M**, **D**, **K**). So the solution is $X_0 = 0_{n\times n}$ and the central controller reads $\widetilde{K}_0 = -C_1$.

Then, for any optimal or suboptimal controller $\widetilde{K}$, we can write:

$$F_l(\widetilde{P}, \widetilde{K}) = F_l(P, \mathbf{F}^{-1}\mathbf{M}\widetilde{K}),$$

so the central controller of the initial problem $P(s)$ reads:

$$K = -\mathbf{F}^{-1}\mathbf{M}C_1 = \mathbf{F}^{-1}[\mathbf{K} - \mathbf{M}\text{diag}(\omega_i^2) \quad \mathbf{D} - \mathbf{M}\text{diag}(2\xi_i\omega_i)] = \mathbf{K_0}.$$

We can recognize the nominal state-feedback defined in equation [3.6]. It can be shown that the LMI-based H_∞ solver applied on the problem $\widetilde{P}$ provides a dynamic controller with a DC gain equal to the central state-feedback $\mathbf{K_0}$ augmented with $2n$ very fast dynamics. These fast eigenvalues can be easily reduced *a posteriori*. Indeed, the LMIs to be solved on the problem $\widetilde{P}$ become marginal and depend only on specifications $(\xi_i,\ \omega_i)$. The optimization problem is then recast in a standard linear problem:

$$\min \mathbf{C}^T\mathbf{X} \text{ subject to LMIs: } L(\mathbf{X}) < R(\mathbf{X})$$

where **X** is the vector of decision variables consisting of:

– γ: the H_∞ performance index;

– the upper part coefficients of symmetric positive matrices X_1 and Y_1 solutions of two H_∞ Riccati inequalities for $\widehat{\gamma}$ (the optimal value).

It can be easily checked that the default values for **C** (in function `hinfsyn`, for instance) are 1 for γ and 10^{-6} for the coefficients of matrices X_1 and Y_1. Such a tuning leads to very fast eigenvalues in the resulting optimal controller.

A3.4. Bibliography

[GAD 06] GADEWADIKAR J., LEWIS F., XIE L., KUCERA V., ABU-KHALAF M., "Parameterization of all stabilizing H_∞ static state-feedback gains: application to output-feedback design", *Proceedings of the 45th IEEE Conference on Decision and Control*, San Diego CA, pp. 3566–3571, December 2006.

[MAG 02] MAGNI J.-F., *Robust Modal Control with a Toolbox for Use with MATLAB*, Springer, 2002.

[ZHO 96] ZHOU K., DOYLE J. C., GLOVER K., *Robust and Optimal Control*, Prentice Hall, 1996.

Appendix 4

Help of Matlab® Functions

A4.1. Function cor2tfg

```
  Observer-based realization of a given controller
 ==============================================================

  [T,F,G] = COR2TFG(PLANT,SYS_K)
  Given an nk-order controller SYS_K defined by a state-space
  realization (Ak, Bk, Ck, Dk) associated with state variable Xk,
  given an n order given plant PLANT defined by a state-space
  realization (A,B,C,D) associated with state variable X,
  COR2TGF computes the matrix T (nk x n) of the linear combination
  Z=TX of the plant states X that are estimated by the controller
  state Xk; that is, Xk=Z_hat.
  F and G are the dynamic matrix and the input matrix of the
  Luenberger observer associated with the controller, that is:
  let u and y be the input and the output of the plant, the
  controller SYS_K can be parametrized in the following way:

    (s or z)X_k = F X_k + G (y-Du) + T B u
              u = Ck X_k + Dk y

  Remark A4.1.- * SYS_K, PLANT are defined as SS objects;
* feedback between PLANT and SYS_K is positive;
* continuous- or discrete-time cases are supported;
* a real solution may not exit.

  This function plots the map of closed-loop eigenvalues (red x)
  and PLANT open-loop eigenvalues (blue +) in the complex plane.
```

```
Then, the user can choose, in an interactive procedure, the
nk closed-loop eigenvalues, which must be assigned to the
Luenberger observer dynamics (spec(F); marked with red o).

The function ensures that a plant uncontrollable eigenvalue
is not assigned to spec(F).
The controller SYS_K is assumed to be minimal.
Auto-conjugate eigenvalues are assigned together.

[T,F,G] = COR2TFG(PLANT,SYS_K,TOL) allows a tolerance TOL
(default: 10^-6) to be taken into account in the uncontrollable
subspace computation.

Reference: Alazard D.,  Apkarian P. "Exact
observer based   structures for arbitrary compensators",
International Journal of Robust and Non-Linear
Control, Vol. 9, no. 2, pp. 101-118, 1999.

See also cor2tfga, cor2obr, cor2obra, obr2cor.
```

A4.2. Function cor2obr

```
  Observer-based realization of a given controller
 ================================================================

  [Kc,Kf,Q] = COR2OBR(PLANT,SYS_K) computes a real observer-based
  realization, that is the Youla parameterization (defined by Kc,
  Kf and Q), of a given controller SYS_K for a given plant
  PLANT in the case:    NK (SYS_K order) >= N (PLANT order).

  In the discrete-time case, PLANT and SYS_K must have
  the same sampling period. Then, the function computes the
  predictor observer-based realization.

  Remark A4.2.- * SYS_K, PLANT and Q are defined as SS object;
* feedback between PLANT and SYS_K is positive;
* a real solution may not exit;
* NQ (order of Q) = NK - N.

  This function plots the map of closed-loop eigenvalues (red x)
  and PLANT open-loop eigenvalues (blue +) in the complex plane.
  Then, the user can choose, in a interactive procedure, the
  closed-loop  eigenvalues distribution between:
```

```
  * state-feedback dynamics [A-BKc] (blue o);
  * state-estimation dynamics [A-KfC] (red o);
  * Youla parameter dynamics (Q)  (green o).

Uncontrollable eigenvalues are automatically assigned to [A-BKc].
Unobservable eigenvalues are automatically assigned to [A-KfC].
(the controller SYS_K is assumed to be minimal).
Auto-conjugate eigenvalues are assigned together.

[KC,KF,Q] = COR2OBR(PLANT,SYS_K,TOL) allows a tolerance TOL
(default: 10^-6) to be taken into account in the unobservable
and uncontrollable subspaces computation.

OBR = COR2OBR(...) creates a structure variable OBR with
fields:
  * OBR.mod : the model of the plant (=PLANT);
  * OBR.Kc  : state-feedback gain;
  * OBR.Kf  : state-estimator gain;
  * OBR.Q   : Youla parameter (ss);
  * OBR.obr : the observer-based realization (ss) of the
              controller;
  * OBR.M   : the transformation matrix between the old and the
              new state-space realizations of the controller:
                  X_k = M [X_hat X_Q].

Reference: Alazard  D., Apkarian P., "Exact
 observer based  structures  for arbitrary compensators",
International Journal of Robust and Non-Linear Control,
 Vol. 9, no. 2, pp. 101-118, 1999.

See also cor2obra, obr2cor, cor2tfg, cor2tfga.
```

A4.3. Function obr2cor

```
SYS_K=OBR2COR(PLANT,K,G,Q)
computes the state-space realization SYS_K=ss(AK,BK,CK,DK)
of the observer-based controller defined by the
Youla parameterization:
   - K: state-feedback gain;
   - G: state-estimation gain;
   - Q: dynamic Youla parameter.
This parameterization concerns the plant defined by PLANT.
```

```
PLANT,Q and SYS_K are defined as SS objects
(continous- and discrete-time cases are considered).

Let us note: [A,B,C,D,Ts]=ssdata(PLANT), then the controller
realization is:
```

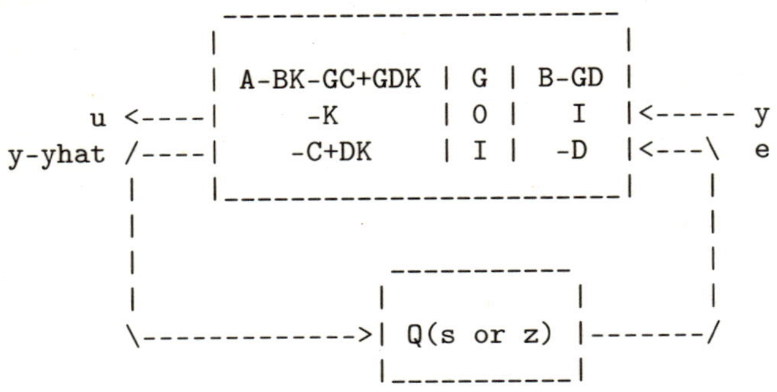


```
Remark A4.3.- positive feedback is assumed for SYS_K.

Reference: Alazard D., Apkarian P.,  "Exact
observer based structures for arbitrary compensators",
International Journal of Robust and Non-Linear Control,
no. 9, pp. 101-118 1999.

See also cor2obr, cor2obra, obr2cor2ddl.
```

A4.4. Function obr2cor2ddl

```
COR=OBR2COR2DDL(PLANT,K,G,Q,H)
computes the state-space realization of the 2 degree-of-freedom
observer-based controller COR defined by the Youla
parameterization:
    - K: state-feedback gain;
    - G: state-estimation gain;
    - Q: dynamic Youla parameter;
  (this parameterization concern the plant defined by PLANT)
    - H: a static feedforward matrix (H=eye(size(K,1)) by default).

Inputs of COR are measurement (y) and reference input (r).

PLANT,Q and COR are defined as SS objects
(continous- and discrete-time cases are considered).
```

Let us note: [A,B,C,D]=ssdata(PLANT), then the controller realization is:

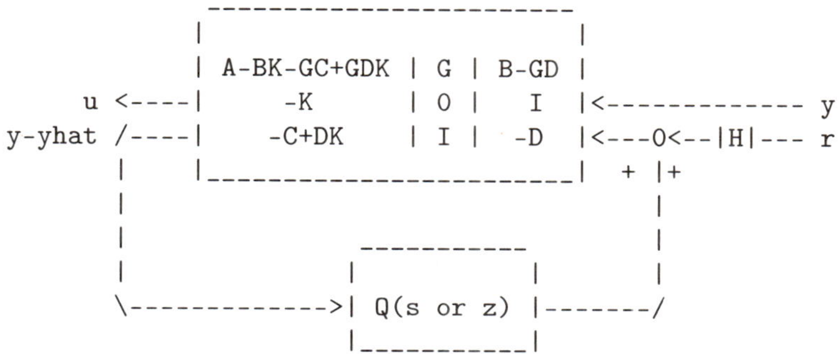

Remark A4.4.- * continuous-time formulation (or predictor form in discrete-time) for the state estimator;
* positive feeedback.

See also cor2obr, cor2obra, obr2cor.

Other syntax:
COR=OBR2COR2DDL(PLANT,K,G,Q,H,M) allows to take into account a static input matrix M=[M1;M2] between H.r and
* xHatDot (or xHat(k+1)) through matrix M1;
* xQdot (or xQ(k+1)) through matrix M2.

A4.5. Function obcanon

Observer-based realization of a given system on a canonic system (with same order).
===

OBR = OBCANON(PLANT) computes a real observer based realization, that is the Youla parameterization (defined by OBR.Kc, OBR.Kf and OBR.Q), of a given (continuous- or discrete-time) system PLANT on a canonical system provided in OBR.mod.

OBR is a structured variable with several fields.

The canonical system OBR.mod is defined by:

```
  OBR.mod=ss([zeros(n,1) [eye(n-1);zeros(1,n-1)]],...
      [zeros(n-p,p);eye(p)],[eye(m) zeros(m,n-m)],zeros(m,p));
```

where n, m and p are the order, the number of imputs and the

```
number of outputs of PLANT, respectively. That is: OBR.mod
(mxp) is a set of n integrators (or delays) in series with:
      * the inputs of the first p intergrators (delays) are
        are connected to the p inputs;
      * the ouputs of the last m intregrators (delays) are
        connected to the m ouputs.

Remark A4.5.- * A minimal state-space realization of PLANT is
                first computed.
 * Auto-conjugate eigenvalues are assigned together.

OBR.obr is the new state-space realization of PLANT.
OBR.M is the transformation matrix between the old and the
      new state-space vector of the system:
                  X_plant = M [X_hat]

See also obr2cor, cor2tfg, cor2tfga, cor2obr, cor2obra.
```

A4.6. Function cor2tfga

```
Observer-based realization of a given controller
==========================================================

[T,F,G] = COR2TFGA(PLANT,SYS_K)
Given an nk-order controller SYS_K defined by a state-space
realization (Ak, Bk, Ck, Dk) associated with state variable Xk,
given an n-order given plant PLANT defined by a state-space
realization (A,B,C,D) associated with state variable X,
COR2TGFA computes the matrix T (nk x n) of the linear combination
Z=TX of the plant states X that are estimated by the controller
state Xk; that is, Xk=Z_hat.
F and G are the dynamic matrix and the input matrix of the
Luenberger observer associated with the controller, that is:
let u and y be the input and the output of the plant, the
controller SYS_K can be parametrized in the following way:

  (s or z)X_k = F X_k + G (y-Du) + T B u
            u = Ck X_k + Dk y

Remark A4.6.- * SYS_K, PLANT are defined as SS objects;
 * feedback between PLANT and SYS_K is positive;
 * continuous- or discrete-time cases are supported;
 * a real solution may not exit.
```

```
Among the combinatory set of solutions T, F and G, a particular
solution is selected. This solution consists of isolating the
n closed-loop eigenvalues located on root loci starting from
the n plant open-loop eigenvalues. The NK other closed-loop
eigenvalues are then assigned to F (and thus does not include
plant uncontrollable eigenvalues).
(This root locus considers the same loop gain varying from 0
to 1 on all control channels.)

The controller SYS_K is assumed to be minimal.
Auto-conjugate eigenvalues are assigned together.

[T,F,G] = COR2TFGA(SYS,COR,NBPOINTS) allows to fix the number
of intermediate points on root loci (200 by default).

[T,F,G] = COR2TFGA(PLANT,SYS_K,NBPOINTS,TOL) allows a
tolerance TOL (default: 10^-6) to be taken into account in the
uncontrollable subspace computation.

See also cor2tfg, cor2obr, cor2obra, obr2cor, obrmap.
```

A4.7. Function cor2obra

```
Observer-based realization of a given controller
================================================================

[Kc,Kf,Q] = COR2OBRA(PLANT,SYS_K) computes a real observer based
realization, that is the Youla parameterization (defined by Kc,
Kf and Q), of a given controller SYS_K for a given plant
PLANT in the case:    NK (SYS_K order) >= N (PLANT order).

In the discrete-time case PLANT and SYS_K must have
the same sampling period. Then, the function computes the
predictor observer-based realization.

Remark A4.7.- * SYS_K, PLANT and Q are defined as SS object;
 * feedback between PLANT and SYS_K is positive;
 * a real solution may not exit;
 * NQ (order of Q) = NK - N.

Among the combinatory set of solutions (Kc,Kf,Q), a particular
solution is selected. This solution consists of assigning to the
```

```
state-feedback dynamics the N closed-loop eigenvalues located
on root loci starting from the N plant open-loop eigenvalues.
(This root locus considers the same loop gain varying from 0
to 1 on all control channels.)
The NK other closed-loop eigenvalues are assigned to A-KfC
and to the Youla parameter Q dynamics.
The proposed assignment can be plotted in the complex plane
using function OBRMAP.

Uncontrollable eigenvalues are automatically assigned to [A-BKc].
Unobservable eigenvalues are automatically assigned to [A-KfC].
(the controller SYS_K is assumed to be minimal).
Auto-conjugate eigenvalues are assigned together.

[KC,KF,Q] = COR2OBRA(SYS,COR,NBPOINTS) allows to fix the number
of intermediate points on root loci (200 by default).

[KC,KF,Q] = COR2OBRA(PLANT,SYS_K,NBPOINTS,TOL) allows a
tolerance TOL (default: 10^-6) to be taken into account in the
unobservable and uncontrollable subspaces computation.

OBR = COR2OBRA(...) creates a structure variable OBR with
fields:
  * OBR.mod : the model of the plant (=PLANT);
  * OBR.Kc  : state-feedback gain;
  * OBR.Kf  : state-estimator gain;
  * OBR.Q   : Youla parameter (ss);
  * OBR.obr : the observer-based realization (ss) of the
              controller;
  * OBR.M   : the transformation matrix between the old and the
              new state-space realizations of the controller:
                  X_k = M [X_hat X_Q].

See also cor2obr, obr2cor, cor2tfg, cor2tfga, obrmap.
```

A4.8. Function obrmap

```
OBRMAP(OBR) plots the closed-loop eigenvalues map
of the  system OBR.mod in positive feedback
with the observer-based controller defined by the
Youla parameterization (OBR.Kc, OBR.Kf,
OBR.Q) associated with OBR.mod:
    - Kc: state-feedback gain,
```

```
    - Kf: state-estimation gain,
    - Q: (ss) dynamic Youla parameter.

The structured input argument OBR can be the output of the
function COR2OBRA (for instance).

On the map are marked:
(considering [A,B,C,D,Ts]=ssdata(OBR.mod))
    - state-feedback dynamics: spec(A-B*Kc) (blue o);
    - state-estimation dynamics: spec(A-Kf*C) (red o);
    - Youla parameter dynamics:  spec(Q.a) (green o).
and also:
    - open-loop plant dynamics: spec(A) (blue +);
    - open-loop controller dynamics: (green +).

Remark A4.8.- * continuous-time formulation (or predictor form in
discrete-time) for the state estimator

[SFDYN,SEDYN,QDYN]=OBRMAP(OBR) does not open a graphics
window but returns:
  * SFDYN: state-feedback dynamics: spec(A-B*K);
  * SEDYN: state-estimation dynamics: spec(A-G*C);
  * QDYN: Youla parameter dynamics spec(Q.a);

See also  cor2obr, cor2obra, cor2tfg, cor2tfga and obr2cor.
```

A4.9. Function h2hinfsol

```
[Kpopt,Kdopt,Copt]=h2hinfsol(m,d,k,w,xi)

Assumption: m>0, xi>0, w>0

H2/Hinfty problem for a second-order mechanical system:
               m*xddot+d*xdot+k*x=u.

Kpopt and Kdopt are the optimal gains of the stabilizing
control law: u=Kp*x+Kd*xdot such that:
  * the acceleration sensitivity function fits the template:
     W(s)=(s^2+2*xi*w*s+w^2)/s^2;
  * the h2norm (J) of the transfer between the distrubance w
     on the acceleration and u is minimized.
Copt is the optimal cost squared (Copt=Jopt^2).
```

```
This function plots also:
 * the frequency-domain responses (sigma) of the open-loop
   and closed-loop acceleration sensitivity function;
 * a 3D plot: J^2=function(Kp,Kd).

Example: [kp,kv,C]=h2hinfsol(10,1,10,0.7,0.6);
```

List of Figures

Index